P9-DTW-755

THE BIOLOGICAL CENTURY

Friday Evening Talks at the Marine Biological Laboratory

Edited by Robert B. Barlow, Jr., John E. Dowling, and Gerald Weissmann

with Pamela L. Clapp

The Marine Biological Laboratory
Woods Hole, Massachusetts

Distributed by Harvard University Press
Cambridge, Massachusetts
London, England

Printed in the United States of America
10 9 8 7 6 5 4 3 2 1

Library of Congress Cataloging-in-Publication Data

The Biological Century: Friday Evening Talks at the Marine Biological Laboratory/edited by Robert Barlow, Jr. ... [*et al.*]
289 p. 000 cm.
Includes bibliographic references and index.
ISBN 0-674-07403-3
1. Biology. 2. Biology—United States. 3. Marine Biological Laboratory (Woods Hole, Mass.)—History. I. Barlow, Robert, 1939- II. Marine Biological Laboratory (Woods Hole, Mass.)

QH311.B49 1993 574—dc20 92-41016 CIP

Book designed by Beth Ready

CONTENTS

CONTENTS

INTRODUCTION

by Garland E. Allen
Washington University

In 1986, Harvard University Press published a volume of ten Friday evening lectures delivered at the Marine Biological Laboratory (MBL) in Woods Hole between 1893 and 1899. Titled *Defining Biology: Lectures from the 1890s* (edited by Dr. Jane Maienschein), these essays present, in the words of the investigators themselves, a portrait of the new American biology that was emerging at the end of the nineteenth century. As the essays reveal, biology was undergoing a major revolution in the 1890s, as it moved from being a largely descriptive and speculative to a more analytical and experimental science. Lengthy anatomical and taxonomic studies were giving way to more physiological and experimental ones. The revolution began largely in embryology, where traditional, descriptive studies were being replaced by a rigorous, causal analysis of differentiation. Heredity, seen in the 1890s as a process governed by hypothetical particles (Weissmann's "ids" and "biophors," for example), was to be replaced a decade later by Mendelian genetics and its highly experimental methodology. It was an exciting period in the history of biology. More importantly, from the MBL's point of view, for the first time Americans were not merely absorbing inspiration from their European colleagues and transmitting it back home; they were also, in many areas, beginning to lead the way.

It is significant that this portrait of biology during this period could be constructed solely from lectures presented at the MBL. If Dr. Maienschein's sampling technique is valid, as I think it is, it is clear that much of the history of the new biology developing at the end of the nineteenth century was centered in one way or another in Woods Hole. Remarkably, the same is still true today. The present volume, a collection of essays based on lectures presented at the same Friday evening series during the MBL's centennial summer of 1988, focuses on some of the most exciting problems in contemporary biology. The lecturers in 1988, like their counterparts a hundred years before, were among the world's leading biologists and represented a wide variety of fields: Gerald Weissmann (biochemistry), Joshua Lederberg (genetics), John B. Gurdon (developmental biology), Shinya Inoué (cell biology), Clifford Slayman (plant physiology), Clay Armstrong (physiology), Meredithe Applebury (molecular biology), Torsten Wiesel (neurobiology), Edward O. Wilson (animal behavior), and John Hobbie (ecology). The history of the Laboratory is thus more than the history of a distinguished institution. It is also the history of biology itself during the past 100 years.

When C. O. Whitman (1842-1910) became the first Director of the MBL in 1888, he took his cue about the laboratory's mission from the highly successful Stazione Zoologica founded fifteen years earlier in Naples, Italy, by German zoologist Anton Dohrn (1840-1909). Dohrn, a protege of Ernst Haeckel (1834-1919) in Jena, Germany, took up the latter's call to study marine organisms because, as the simplest of all living forms, Haeckel thought that they illustrated clearly the most fundamental principles of life. Haeckel's commitment to the theory of recapitulation (expressed in his own unique form as the biogenetic law, or "ontogeny recapitulates phylogeny") meant that he saw marine organisms as earlier evolutionary stages of more advanced forms, and hence the ones that displayed in pristine clarity the fundamental processes characteristic of all living things. The Naples Station, opened in 1873 to exploit the rich marine fauna and flora of the Bay of Naples, proved that Haeckel's vision was a fruitful one, if for reasons other than those he had supposed.

The Stazione became a mecca for biologists the world over during the last quarter of the nineteenth century. Whitman, who had obtained his Ph.D. under Rudolf Leuckart in Leipzig, Germany, had worked at the station in its earliest years, first in 1875 and again in 1881-82. He was soon followed by other Americans: Edmund Beecher Wilson (1856-1939) in 1883 and again in 1892, William Morton Wheeler (1865-1937) and George H. Parker (1864-1955) in 1893, and Thomas Hunt Morgan (1866-1945) in 1894-95. All were to become mainstays of the MBL from the outset, and all are among those honored in the lectures included in this centennial volume.

The MBL was in many ways, both directly and indirectly, an outgrowth of the Naples Station. Directly, Whitman's experience at the station as a young investigator and also as a personal friend of Anton Dohrn taught him much about the value of marine organisms for study and the benefits of a community of investigators coming together to carry out research on common problems. Indirectly, the same excitement about a "new" experimental biology that fostered the development of the Naples Station also inspired Whitman and was incorporated into the structure of the MBL. For a century now, the Stazione Zoologica and the MBL have been in constant association and cooperation: the Stazione hosted the MBL's initial centennial planning session in October, 1984; the Director of the Stazione regularly has a summer lab at the MBL, and the two institutions cooperate in offering alternate-year courses on the history of the life sciences.

What characterizes the work carried out at the MBL from the outset was the organization of research around biological problems, not specific organisms and their taxonomy. To be sure, taxonomic work has been carried out at the MBL from the first summer down to the present, but almost always centered on a problem of larger scope than mere categorization. Darwin's *Origin of Species* had shown, perhaps more clearly than any other work on natural history in the nineteenth century, that the study of life and its history involved a number of problem areas: heredity (what was inherited and in what patterns); development (the meaning of recapitulation or how differentiation is guided); phylogeny (how historical relationships are inferred); ecology (how relationships between organisms and environment are established); and physiology (how function is related to structure as an aspect of adaptation). These problems all emerged in a Darwinian context, and were thus initially dominated by evolutionary questions. By the time the MBL was founded, however, new interests were already being expressed by a younger group of workers. Problems such as the nature of heredity or embryonic development could be studied for their own sake, not only for the light they could throw on evolutionary questions. We need only reflect on the fact that Hans Driesch (1867-1941) was a year-round investigator at the Stazione at the time of his famous controversy with Wilhelm Roux (1850-1924) over mosaic *versus* regulative development, and that the young T. H. Morgan spent a year (1894-95) working with Driesch at Naples, to recognize how much the issue of pursuing biological problems, rather than descriptive morphology, was intertwined with the early history of marine stations.

Key to the excitement felt by many of the young Americans who went to Naples and who later returned year after year to the MBL was not only the investigation of biological "problems," but also the mission of making biology a more experimental, quantitative, and rigorous science, in the mold of physics and chemistry. This new biology at the MBL was thus dominated from the beginning by physico-chemical reductionism, or the "mechanistic philosophy of life," a phrase coined and made so memorable by MBLer Jacques Loeb (1858-1926) in his essay of the same name in 1911. The MBL grew up with the spirit of life as a physico-chemical system, understandable as the sum of individual chemical processes. The problem-oriented and experimental science that was part of the MBL's research has remained with it to the present day.

The dialectic of continuity and change characterizes the history of any institution or community of people over time. The MBL is no exception. A look at the topics for the Friday evening lectures between 1890 and 1899 (a complete list is reprinted in Maienschein's *Defining Biology*, pp. 51-56), and at the table of contents of the present volume, shows some remarkable similarities as well as some unexpected differences. Particularly striking are the recurrent, or perhaps continuing, problems that appear at opposite ends of the century. From the outset, the problem of embryonic development has dominated much of MBL's research. This is partly a practical matter, of course, because marine eggs and embryos are particularly easy to handle in the laboratory. But it also reflects the fact that the problems of development have been, from the 1890s onward, so alluring. The names of E. G. Conklin (1863-1952), Ross G. Harrison (1870-1959), Frank R. Lillie (1870-1947), C. M. Child (1869-1954), and Viktor Hamburger (b. 1900)—all MBL regulars up to the mid-century—attest to the importance of this subject at the laboratory. A close runner-up is the study of cell division and related problems of cell structure and function. From Shosaburo Watase's pioneering studies on karyokinesis and the role of the centrosome in 1894 to studies on cell organelles by Keith Porter in the 1960s and 1970s, to those on the cytoskeleton and mitotic spindle by Shinya Inoué and Robert Palazzo today, the fine structure of the cell has occupied the attention of generations of MBL investigators.

Another topic that has transcended the years with continued intensity is the study of behavior. Beginning with Whitman himself, whose later research work was devoted to pigeons, through his younger colleagues William Morton Wheeler, Jacques Loeb, and Herbert Spencer Jennings (all three of whom gave lectures on behavior in the 1890s), to Edward O. Wilson today, the study of animal behavior has always been a part of the MBL research tradition. Ecology has also had its continuing advocates, especially the study of marine ecosystems, given a great boost after the founding of the Woods Hole Oceanographic Institution (WHOI) in 1931. The names of Spencer Fullerton Baird (1823-1887), founder of the U. S. Fish Commission and its laboratory in Woods Hole in 1882, H. P. Bigelow (1879-1967), first Director of WHOI, and Alfred Redfield (1890-1983), MBL investigator and trustee and Associate Director of WHOI from 1924-1957, are all associated with the attempt to make study of the oceans and large ecosystems amenable to quantitative, exact methods.

Among the differences between the 1890s and the 1980s, however, are some surprises. Topics of evolutionary theory dominated many of the lectures between 1890 and 1899. John A. Ryder (1852-1895) and Henry Fairfield Osborn (1857-1935), both strong anti-Darwinians, gave two lectures each in those summers, along with less-well-known figures such as paleontologist William Berryman Scott and ornithologist Herman C. Bumpus. By contrast, the topic of evolution *per se* received no lecture during the centennial year. I interpret this to mean that the concept of evolution by natural selection is so much a part of modern biology that it did not need to be brought forward and either defended or attacked as a paradigm as was the case in the 1890s. It is also the case that, since the early years of the century, the MBL has concentrated largely on problems of proximal, as opposed to ultimate causation: that is, on problems of function rather than of origin. An evolutionary paradigm underlies most research at MBL today in a far more solid way than it did through much of the 1890s, but research on the mechanisms of evolution itself, and on major issues in modern evolutionary theory, have not been prominent at MBL for many decades. This situation is being rectified in part, however, by the recent addition of a year-round laboratory and program on molecular evolution, headed by Mitchell Sogin.

In the opposite direction, neurobiology (an anachronism, of course, applied to the period before Stephen Kuffler coined the term in 1959) received only scant attention in the 1890s (two lectures on the function of the ear, and one each on nerves and muscles), but was represented by three lectures (out of 10) during the centennial season. Clearly, the study of nerves and sensory systems has expanded by leaps and bounds in the intervening years. Another difference is that in the 1890s a dominating interest was morphology, that multidisciplinary field aimed at elucidating phylogenetic relationships by using comparative anatomy, physiology, ecology, and embryology. A largely descriptive and highly speculative research program, morphology gradually gave way in the early decades of the twentieth century to experimental and analytical studies. No lectures in the centennial summer dealt with morphology as such.

Finally, in the 1890s a number of lectures dealt explicitly with the "new" methods of experimental, quantitative, and analytical work that represented the "cutting edge" of biology at the time (I count six lectures on methodology between 1890 and 1899, including several by Loeb and Whitman, along with William Morton Wheeler's translation of the introduction to volume 1 of the *Archiv für Entwicklungsmechanik*, in which founder and editor Wilhelm Roux spells out the new causal-analytical approach to development). By contrast, in the centennial summer only one lecture, Gerald Weissmann's essay on Loeb, dealt at all with such methodological issues. This implies that the analytical and physico-chemical method that was new and innovative in 1890, is so taken for granted today that no one feels the necessity of even discussing it openly (most philosophers of science would find this a dangerous situation, on the principle that no philosophy is more likely to lead astray than one that is unexamined).

As is evident from an inspection of the table of contents of this volume, each centennial lecture is dedicated to the work of a former MBL pioneer in one or another field. Neurobiology lectures are dedicated to K. C. Cole, A. L. Hodgkin, H. K. Hartline, and Stephen Kuffler; while a lecture on the biochemistry of vision is dedicated to Selig Hecht and George Wald; a lecture on genetics is dedicated to Thomas Hunt Morgan and A. H. Sturtevant; one on animal behavior to C. O. Whitman and William Morton Wheeler; on the role of ions in plant cell function to W. J. V. Osterhout; on cell fine structure to Keith R. Porter; on ecology to Spencer Baird, H. P. Bigelow, and Alfred Redfield; and on developmental biology to Edwin Grant Conklin and Frank R. Lillie. A lecture on the mechanistic and reductionistic view of life is dedicated to Jacques Loeb, while a special essay (not given as a lecture in the Friday evening series, but which is based on a two-day centennial symposium in his honor) is dedicated to, and focuses on, the work of the late Albert Szent-Györgyi.

The general format of each lecture is preserved throughout the volume: there is an historical introduction by a member of the MBL community that focuses on the work of the figure being honored, followed by a brief introduction to the lecturer. Some of the introductions are purely biographical, such as Robert Goldman's outline of Keith Porter's career; others, such as Franklin Offner's survey of neuron recording techniques from K. C. Cole to A. L. Hodgkin and H. E. Huxley, more broadly trace the history of a field. Still others are more strictly historical in scope, as is Jane Maienschein's discussion of Jacques Loeb's pursuit of a physico-chemical view of life.

The essays themselves are almost all some mix of history and current science. On the historical end are Joshua Lederberg's tracing of the process of genetic mapping in *Drosophila* from Morgan and Sturtevant in 1911, through his own work with E. L. Tatum on mapping the chromosome of *E. coli* in the 1940s, through Seymour Benzer's mapping of phage T-4 in the 1960s. In the same vein is John B. Gurdon's tracing of the problem of cellular determinants of development from E. G. Conklin's earliest studies on the role of the "yellow plasm" in the cleavage of ascidian eggs in the late 1890s, through the work of William Jeffrey's elucidation in the 1980s of the role of calcium ions in localizing the "yellow substance" within the egg. On the other end of the spectrum, Torsten Wiesel's essay focuses more on current research into the interpretation of retinal information, building on the work of H. K. Hartline and Stephen Kuffler in the 1940s and 1950s.

Two essays in this collection deserve special mention. One is Benjamin Kaminer's biographical/scientific account of the work of Albert Szent-Györgyi, who died the year prior to the centennial at the age of 94. Kaminer, a many-times collaborator with Szent-Györgyi over the years, captures well the iconoclastic, tenacious, sometimes outrageous, but always humorous and insightful nature of "Prof," who graced the MBL year-round for more than three decades. Probably the most broad-ranging and imaginative of the lectures is Gerald Weissmann's comparison of Jacques Loeb and Gertrude Stein—an unlikely-seeming pair if there ever was one, but bound together, as Weissmann sees it, by their common assent to mechanistic materialism. Stein's involvement in the MBL centennial is not purely spurious, however. She was a student in the embryology course in 1897, just prior to her matriculation at Johns Hopkins Medical School, and at a time when Jacques Loeb was beginning to frequent the lab in Old Main that would be his summer research haven for years to come. One learns perhaps most about the ambiance of MBL from these essays, both of which emphasize so vividly the personal, social and intellectual side of life at the laboratory.

This volume is not a systematic history of the MBL during its first one-hundred years, nor is it a chronicle of the development of biology during that time. Rather, it presents a picture of current research problems, most with a very long tradition, that have always occupied investigators at the laboratory. There is a lot of history and a lot of science in these pages, and it is a happy, if eclectic combination. Reading through the essays and their historical introductions should provide any reader, whether biologist, historian, or simply interested layperson, with some sense of what the MBL is, and why it has been so important in the development of biology in the twentieth century. Most of all, these lectures present in a way very few such volumes do, a portrait of the human side of science—the excitement, the mistaken paths, the interaction of personal and social factors that shape research traditions and, indeed, of the way we come to view the world around us.

Jacques Loeb
(1859 - 1924)

The Mechanistic Conception of Life: Loeb the Teacher, Stein the Student at the MBL

Introduction by Jane Maienschein

Let's go back for a moment to one hundred some years ago, when a visitor to the MBL in the 1890s reported walking through deserted buildings and empty laboratories. He was curious about the place and its people, but unfortunately it was a Sunday morning and nobody was around. He found only a set of simple gray wooden structures, strung together as wings of the same building. He walked through the halls and heard only the sound of his own footsteps—until he went upstairs and paused outside one open door. The small energetic man inside called the visitor in, urged him to peer through his microscope, and excitedly discussed (in a thick German accent) the discoveries of the day. The visitor knew then that he had found the essence of the MBL in the enthusiastic, energetic, obvious love of science that this scientist exuded. The scientist, of course, was **Jacques Loeb**.

The MBL does involve the excitement and love of doing science that the visitor witnessed. But it includes something more. The MBL also means the sharing of science and the enthusiasm for the chase, so to speak, in the form of public education about the processes of doing science. The MBL has always involved the combination of doing science and the open, public discussion of ideas and their consequences. Loeb's eagerness to share his work with the stranger represents both aspects of what the MBL was—and has continued for over one hundred years to be—about. It is particularly appropriate that Loeb was honored on the MBL's Dedication Day, because Loeb saw the public side as well as the research side of science especially keenly.

Loeb wanted more than to gather knowledge or to accumulate facts. He wanted above all to use knowledge to control life. (See Philip Pauly's excellent *Controlling Life: Jacques Loeb and the Engineering Ideal in Biology*, Oxford University Press, NY. 1987.

Paper edition with University of California Press.) At first exploring problems of human and brain physiology, the German researcher had moved for financial reasons to ophthalmology. He quickly gave that up, however, and turned to marine studies and to "physiological morphology" during research visits at the Naples Zoological Station. After marrying an American, he resolved to move to the United States to pursue a scientific rather than a medical career. Bryn Mawr College had an opening because E. B. Wilson and physiologist F. S. Lee had just left for Columbia. T. H. Morgan had agreed to serve as the head of biology, but Bryn Mawr needed a physiologist as well. Unfortunately, Loeb was a German Jew—a problematic status indeed. Nonetheless, he was hired in 1891, but stayed only one year before moving on to the University of Chicago. His very first summer, Loeb also accepted C. O. Whitman's invitation to join the MBL staff as the head of a new physiology course, a position from which he inspired considerable interest in that functional side of biological processes.

Loeb's successful production of artificial parthenogenesis in sea urchins provoked great public attention, with newspaper headlines screaming about "Creation of Life" and "Immaculate Conception Explained." Public acclaim and controversy followed. In fact, other researchers, including Morgan, had developed techniques for initiating the division of sea urchin eggs without male fertilization. But only Loeb saw why that phenomenon might prove interesting. And only Loeb worked at getting the physico-chemical conditions just right so that he could make the development following artificial parthenogenesis work. Science should gain knowledge and understanding, then use it to make life better, he insisted.

In 1910, Loeb moved from the University of California (where he had gone from Chicago) to a research position at the Rockefeller Institute for Medical Research. With a lab provided for him at the MBL in addition, Loeb was free to do his research without outside pressures. In this role, he became the model for Sinclair Lewis's pure scientist Max Gottlieb in *Arrowsmith*. But Loeb remained concerned with "engineering" or control problems as well. He argued that man's ethical system, for example, must be based in evolution-based instincts and therefore ultimately depends on chemistry and heredity. Similar understanding of other life phenomena would follow from a proper "mechanistic conception of life." With understanding would come guidance and control.

With such a set of emphases, Loeb provided an important influence at the MBL and for American science more generally. His enthusiasm for science and for research inspired many—even many who disagreed with his particular approach, with his conclusions, or with his personal ideals. His concern with the public side of science extended his influence.

It is worth recalling the original purpose of the MBL to see how well Loeb fit in. For like Loeb, the MBL had a public and educational goal for science. The MBL began when members of the Boston Society of Natural History and the Woman's Education Association of Boston decided to establish a place for teachers to teach biology to teachers. Massachusetts had made this necessary by passing a law requiring that science, including biological science, be taught in the public schools. Louis Agassiz's short-lived Penikese Island School and Alpheus Hyatt's Annisquam School had responded to the challenge, but neither was intended to be permanent. Massachusetts needed a permanent place to combine the doing of biological science with the teaching of that science, both to carry out new work and to carry the work to the public—all as part of the move to progressivism that intended to use science to improve the world.

The first group at the MBL was half women; the supporters included prominent women; and soon a few women even became instructors. In fact, women had become active educators in biology at a time when the women's colleges led the way in science teaching. Remember that Wilson, Morgan, and Loeb, for example, all really began their professional careers at Bryn Mawr College. Women have remained a central part of the MBL in various ways for one hundred years, and so has the public, educational mission of the place: through courses, through the Friday Evening Lectures, through other lecture series, through special events such as the MBL's one hundreth birthday celebration, and, above all, through the work of scientists such as Jacques Loeb and Gerald Weissmann.

Weissmann's recent volumes of essays about medical practice and medical science have reached a wide audience, an audience eager to learn not just the facts of science or the great discoveries but also about the process of scientific discovery, about the excitement of doing science, of succeeding and of failing. People want to learn about the social and historical aspects shaping scientific work. They want to understand and to be able to use science. Many want to learn in order to be better able to teach about science themselves. Gerald

Weissmann cares about addressing these concerns as he cares about the public and educational goals of science. He perpetuates the spirit of the MBL as Loeb did. It is therefore particularly appropriate that Gerald Weissmann honored Jacques Loeb on the one-hundred-year anniversary of the dedication of the Marine Biological Laboratory.

Gerald Weissmann is a graduate of Columbia College (1950) and NYU School of Medicine (1954). After clinical training at Mt. Sinai and Bellevue Hospitals (under Lewis Thomas) he did postdoctoral work in the Department of Biochemistry at NYU (with Severo Ochoa) and at the Strangeways Research Laboratory, Cambridge (with Dame Honor Fell). Since then, he has remained at NYU's Department of Medicine, where he has been a professor of medicine and director of the Division of Rheumatology since 1973. His research on inflammation and the structure of lipid membranes (liposomes) has won him the Robecchi Prize in Rheumatology, the Lila Gruber Award for Cancer Research, and the Distinguished Investigator Award of the American College of Rheumatology (1992). He has been president of the American Rheumatism Association (1982-83) and the Harvey Society (1981-82). He has spent every summer since 1970 doing research at the MBL; he received the MBL award in 1976 and 1979. An essayist for *Hospital Practice* magazine since 1973, Dr. Weissmann's essays and reviews have appeared in *The New York Times*, *The Washington Post,* and *The New Republic*. They have been collected into three volumes: *The Woods Hole Cantata* (Dodd, Mead; 1985), *They All Laughed at Christopher Columbus* (Times; 1987), and *The Doctor with Two Heads* (Knopf; 1990). Dr. Weissmann is currently the editor of *MD* magazine.

THE MECHANISTIC CONCEPTION OF LIFE: LOEB THE TEACHER, STEIN THE STUDENT AT THE MBL

GERALD WEISSMANN

New York University Medical Center

FIGURE 1 SHOWS A SCORE OF STUDENTS in the summer dress of a century ago collecting specimens at low tide from a harbor near Woods Hole in Massachusetts. The harbor is Quissett, and its waters today remain rich in marine life; its heights are still dominated by a Yankee cottage called "Petrel's Rest"; the house is still surrounded on four sides by a veranda and fronted by a flagpole. The young people are on a collecting trip for the course in invertebrate zoology at the Marine Biological Laboratory; its students still collect

Figure 1. *MBL 1897 Invertebrate Course collecting trip. Gertrude Stein is at center in the foreground.*

specimens from the inlet of Buzzard Bay. The photo was taken on July 31, 1897: the young woman in the middle is Gertrude Stein. She has turned, smiling to her brother Leo, who holds up a specimen jar: "Look what I've found!" he gestures to the photographer. Many in the group are also smiling, it is the height of summer; they are young and have disembarked at a marine Cythera where lush creatures drift on the tide. Leo Stein has snared a ctenophore, a solitary free swimmer that resembles a jellyfish.

For a century now, images of the Marine Biological Laboratory have been kept like wallet snapshots in the memories of scientists who return there yearly and of their students who do not. Those summer snapshots do not evoke memories of a routine, workaday research institute. They conjure up spirits of a blithe community that springs to life each year—like Brigadoon—for experiments by a northern sea warmed by Gulf stream currents. Not all students of the MBL become scientists, some may not even learn much science, but few will forget those moments at the beach when the cry goes up: "Look what I've found!"

At Woods Hole, Gertrude Stein was 23; she had just finished Radcliffe and was to enter the Johns Hopkins Medical School in the fall. That summer she had enrolled in the embryology course at the MBL and often accompanied Leo and the invertebrate zoologists on collecting trips. W. C. Curtis, who is shown in the photograph in white cap and knee-boots just to the right of Stein, remembered in the *Falmouth Enterprise* (Curtis, 1955): "For us that summer she was just a big fat girl waddling around the laboratory and hoisting herself in and out of the row-boats on collecting trips." Adding incest to injury, he might have added that she was in thrall to her fast-talking brother: some might read the photo as a record of the intense bond. Stein has written that as children she and Leo had tramped alone in the woods of northern California; her adoring gaze in the Quissett photo suggests an American future for the both of them: from the redwood forests to the Gulf stream waters this land was made for you and me.

But the photo suggests motifs other than those of a family romance; the image alone is a stunning icon of natural science. The anonymous photographer has snapped a tableau of figures in a landscape so arranged that all the compositional lines—from the hillside on the upper left to beachgrass at the lower right—converge on Leo who holds high the collecting jar with his discovery. The creature from the ocean has been plonked into a glass pot to become a specimen for science. The composition has elements of Joseph Wright's neo-classic "Experiments with the Air Pump" in that the figures are so disposed as to lead us to the apex of a visual pyramid at which life has been caught in a jug. The photo leads our eye, via the sight-lines of Gertrude and another young woman, straight to Leo who looks to the lens with his prize held high. Looking at the photo almost a century later, we know that the scales of discovery tipped toward the sister, and therefore we tend to read the picture as an action shot of the artist as a young woman. Caught there on the beach at Quissett, she is forever happy by the summer sea from which smiling students fetch treasures for the lab.

No other picture of Stein shows her quite as perky as on that day at Quissett. The group photo of her embryology course, taken later in the summer (Fig. 2), shows her preoccupied and unsmiling. Indeed, nothing about science or medicine seems to have given her much joy after her tussle with ctenophores. After indifferent attention to laboratory and clinic, she failed to graduate from the Johns Hopkins Medical School with her class of 1901. Although she had completed the bulk of her work, she seemed to have floundered over obstetrics—she who had forgotten all that marine embryology. She tells us that it was in the course of her obstetrical work that she became "aware of

Figure 2. *MBL 1897 Embryology Course class photograph. Gertrude Stein is in the front row, second from the left.*

the Negroes" in Baltimore clinics serviced by Johns Hopkins; from that experience emerged the story of Melanctha in *Three Lives*. The mulatto abandons a humdrum doctor in favor of more louche companionship.

Gertrude Stein and Johns Hopkins separated as if by mutual consent. In *The Autobiography of Alice B. Toklas* (1933) she tells us: "The Professor who has flunked her asked her to come to see him. She did. He said, of course, Miss Stein all you have to do is to take a summer course here and in the fall naturally you will take your degree. But not at all, said Gertrude Stein, you have no idea how grateful I am to you. I have so much inertia and so little initiative that very possibly if you had not kept me from taking my degree I would have, well, not taken to the practice of medicine, but at any rate to pathological psychology and you don't know how little I like pathological psychology, and how all medicine bores me. The professor was completely taken aback and that was the end of the Medical education of Gertrude Stein."

Whether from boredom with medicine or deeper battles of the self, no joy shows in her picture with the class of 1901 at Hopkins (Fig. 3). She stands glumly half-hidden in the back row behind other women and the shorter, swarthier of the men. She left for Europe, and after aimless wanderings with her brother, settled in Paris to find her own vocation. Gertrude Stein soon outdistanced Leo who had stopped dabbling in biology; she discovered new art and new friends on the banks of the Seine. By the time Picasso painted her in 1906, her Iberian portrait showed the confident young writer who was crafting *Three Lives* in the course of changing forever the way we hear words.

Those who look to details of biography for "explanations" of literary or artistic styles can usually extract as much material as needed to convince. In the case of Gertrude Stein, the most convincing "explanation" of her unique style was offered by B. F. Skinner (1935), writing in *The Atlantic* of 1935. In an article entitled "Has Gertrude Stein a Secret?" the Harvard professor of psychology traced Stein's technique to her undergraduate research work on automatic writing with William James, Skinner's predecessor in experimental psychology. The case is persuasive, even if Skinner's aesthetic judgments are not. But on a July day at Woods Hole, with ctenophores and jellyfish awash on the tide, another set of influences on Stein seems just as likely. My hunch is that Gertrude Stein was an old-fashioned Woods Hole mechanist, a reductionist of the school of Jacques Loeb. Her revolution of words owed as much to *The*

Figure 3. *Johns Hopkins Medical School class portrait (1901) with Gertrude Stein, top row, third from the right.*

Mechanistic Conception of Life of Loeb (1965 reprint) as to her study of "Normal Motor Autonomism" (Solomons and Stein, 1896), which she had written for William James, or to the "Demoiselles d'Avignon" of her Picasso years (the first versions of which showed a medical student as spectator).

During the summer of 1988, the Marine Biological Laboratory celebrated its centenary, and the Dedication Day lecture was given in honor of Jacques Loeb (1859-1924). Loeb was the leader of the new, mechanistic school of American biology the adherents of which tried to explain the phenomena of biology by the equations of physics and not the quirky logic of vitalism. In 1897, Loeb was teaching the physiology course at the MBL and the implications of his biophysical approach were the talk of the laboratory. He had demonstrated that the chemical nature of salts in the environment of a cell controlled its irritability, movement, and reproduction in predictable ways. He was on his way to creating life in a dish, to forming fatherless sea urchins by chemical means. Parthenogenesis was announced two years later, but his work on tripisms and salt solutions had already paved the way for what was to Loeb the fundamental task of physiology: "to determine whether or not we shall be able to produce living matter artificially."

Loeb and his school of mechanists believed that they were the legitimate heirs of the *philosophes*, and Loeb's book *The Organism as a Whole* (1916) is dedicated to Denis Diderot in the words of John Morley's tribute to the *philosophe*:

"He was one of those simple, disinterested, and intellectually sterling workers to whom their personality is as nothing in the presence of the vast subjects that engage the thoughts of their lives."

The mechanist's credo, with its belief that it is possible to frame disinterested thoughts unshaped by "personality," has been dismissed as shallow reductionism by three generations of philosophers, poets, and divines. But, to paraphrase John Sayles, just because they argue sweeter doesn't mean they're right. Reductionist principles in science have permitted us to conquer tuberculosis, syphilis, polio, and scarlet fever. Using mechanistic models, we have uncoiled the human genome and scanned the living brain; we've tamed mad minds with lithium and patched sick hearts with teflon. We have fulfilled Loeb's prophecy of "producing mutations by physico-chemical means and nuclear material which acts as a ferment for its own synthesis and thus reproduces itself."

But Loeb claimed more territory for science than that of the body. He extended its empire to the mind. Replying to William James' request for his views of brain function, Loeb responded (quoted in Pauly, 1987):

"Whatever appear to us as innervations, sensations, psychic phenomena, as they are called, I seek to conceive through reducing them—in the sense of modern physics—to the molecular or atomic structure of the protoplasm, which acts in a way that is similar to (for example) the molecular structure of the parts of an optically active crystal." (1888)

Loeb detailed this reductionist proposal to the one man perhaps least likely to be persuaded. William James, physician, professor of anatomy before he became a professor of psychology, had argued eloquently the opposite and more generally held view before an audience of Unitarian ministers at Princeton in a lecture entitled "Theism and the Reflex Arc." In a ringing defense of theism against the reductionists of the reflex arc, he told the liberal clergy that (reprinted in James, 1974):

"Certain of our positivists keep chiming to us, that, amid the wreck of every other god and idol, one divinity still stands upright—that his name is Scientific Truth... But they are deluded. They have simply chosen from the entire set of propensities at their command those that were certain to construct, out of the materials given, the leanest, lowest, aridest result,—namely the bare molecular world,—and they have sacrificed all the rest... The scientific conception of the world as an army of molecules gratifies this appetite [for parsimony] after its fashion most exquisitely."

The dialogue between those who lean towards Loeb and those who tend to agree with James has been carried on by the waters of Vineyard Sound and Buzzards Bay for a century. Mechanists maintain that the control of living things must precede our understanding of them, holists contend that when we understand them we may not hanker after control. But Woods Hole has always had room for both parties in this dialogue, as it has for those who come to verbal blows over the conflict between nature and nurture, structure and function, Red Sox and Yankees.

Stein got a whiff of both sides in the course of her research career. She heard the siren song of vital forces from the voice of gentle James; she worked with one of Loeb's closest colleagues, Franklin Pierce Mall, not only in the embryology course at the MBL, but also at Johns Hopkins. But by the time she published *Three Lives*, her critics got it just right: she was on the side of the mechanists. Wyndham Lewis—no great friend of poor folk—argued in his review that Stein's book put demotic speech into the "metre of an obsessing time" and although "undoubtably intended as an epic contribution to the present mass-democracy" gave "to the life it patronizes the mechanical bias of its creator."

My hunch that Gertrude Stein's revolution in words is based on Loeb's mechanistic conception of life derives not only from a re-examination of B. F. Skinner's hypothesis, but also from trends in our culture that became more prominent after his critique. The best biography of Stein, *The Third Rose* by John Malcolm Brinnin (1959), contains no reference to Loeb, nor to the Marine Biological Laboratory. The best biography of Loeb, *Controlling Life: Jacques Loeb and the Engineering Ideal in Biology* by Philip J. Pauly (1987), does not mention Stein. Both books discuss with varying degrees of insight the interaction of their principals with William James, Franklin Pierce Mall, and B. F. Skinner; both books tell stories that contain substantial parallels.

Stein and Loeb were both orphaned as adolescents and brought up by well-off relatives as secular Jews, Stein in Oakland and

Baltimore, Loeb in the Rhineland and Berlin. Both emigrated as young adults, both spoke their adopted languages with awkward accents. Both performed experiments on brain function early in their career, attracting the attention of William James; indeed Loeb wrote to James fishing for a job in the new world. When Stein entered Hopkins, she worked on mechanical models of spinal tracts with Frank Mall, Loeb's colleague from Woods Hole and the University of Chicago. Finally, both engaged the far from casual interest of B. F. Skinner.

Pauly describes how Skinner was persuaded by Loeb's argument as spelled out in *The Organism as a Whole* that mechanist principles could be applied to the study and control of behavior. Skinner was also captivated by *Arrowsmith* (Lewis, 1924), in which Loeb—as Martin Arrowsmith's mentor, Dr. Gottlieb—is depicted as a secular saint of science. Pauly correctly identifies Skinner and the other behaviorists of T-maze and pigeon-box as the rightful heirs to Loeb's mechanistic conception of life. If the founder of experimental psychology at Harvard, William James, had championed the cause of vitalism against the raiders of the reflex arc, his successor was prepared to work for the opposition. B. F. Skinner found support for his views in the work of James' most famous student.

Skinner was the first to draw attention to Gertrude Stein's undergraduate work on automatic writing. He pointed out the origins of her verbal experiments by unearthing the paper on "Normal Motor Autonomism" published in the *Psychological Review* of September 1896 by Leon M. Solomons and Gertrude Stein from the Harvard Psychological Laboratory. (What a pride of lions and Leos! Loeb's not undistinguished brother, a pioneer in studies of inflammation, was also named Leo.)

Solomons and Stein report on experiments designed to test whether a second personality—as displayed in cases of hysteria—could be called forth deliberately from normal subjects. The two authors undertook to see how far they could "split" their personalities by eliciting automatic writing under a variety of test conditions. They concluded that hysteria is a "*disease* of the *attention*" (their italics), basing their argument on the finding that when distracted or inattentive, normal subjects show the abnormal motor behavior of hysterics. It may be no coincidence that Solomons and Stein performed laboratory work on hysteria in the very year that Sigmund Freud and Josef Breuer published their clinical "Studies in Hysteria" (1895). The subject was much in the air on both sides of the Atlantic, not least because—as William J. McGrath has pointed out (*New York Review of Books,* August 18, 1988)—the study of hysteria offered science an opportunity to strike at the foundations of religion. Explain away divine madness by the reflex arc and you explain away divinity. James as a theist was persuaded that experimental psychology would validate all the varieties of religious experience. His students, on the other hand, came to a behavioral conclusion that favored the reflex arc: wrote Solomons and Stein, "An hysterical anesthesia or paralysis is simply an inability to attend to sensations from this part."

Skinner showed little interest in Solomon's and Stein's discussion of hysteria, however congenial to the mechanistic conception it may have been. He was more concerned with tracing Stein's literary style to her experiences of automatic writing. Using themselves as test subjects, Solomons and Stein were able to show that with a little practice they could regularly produce automatic writing as they took dictation while reading another text: "The word is written or half-written before the subject knows anything about it, or perhaps he never knows anything about it. For overcoming this habit of attention we found constant repetition of one word of great value."

After they had succeeded in training themselves by this sort of cognitive drill, and after sessions with ouija boards to call up their alter egos, automatic writing became easy. Stein found it convenient to read what her arm wrote, but following it three or four words behind her pencil. In this fashion, "a phrase would seem to get into the head and keep repeating itself at every opportunity, and hang over from day to day even. The stuff written was grammatical, and the words and phrases fitted together all right, but there was not much connected thought."

Skinner—a traditionalist with respect to the arts—argued that these experiments explained why there appeared to be two Gertrude Steins. The first Stein was accessible, and had written such serious work as *Three Lives* and *The Autobiography of Alice B. Toklas*; the other Stein was dense, and wrote stuff that was grammatical, with words and phrases fitted together all right but without connected thought. The second Stein had written *Tender Buttons*, her portraits, and *The Making of Americans. Four Saints in Three Acts* was yet to come! Skinner gave mild positive reinforcement to the first Stein, but strong negative reinforcement to the second, chiding her that "the mere generation of the effects of repetition and surprise is not in itself a literary achievement." Skinner complained that the second Stein gives no clue as to the personal history or cultural background of the author and dismissed her most adventurous book with a phrase from their common master, William James: "*Tender Buttons* is the stream of consciousness of a woman without a past."

On the surface, Skinner seems to have scored a point. It is easy to pick up resonances between the samples of automatic writing that Solomons and Stein present and the stuff of Stein's later work. Thus the first two passages below, examples of automatic writing from the 1896 article *sound* like the second passages from later Stein (Solomons and Stein, 1896; Stein, 1972); a closer look will show all the difference:

1. "This long time when he did this best time, and he could thus have been bound, and in this long time, when he could be this to first use of this long time..." (Solomons and Stein)

2. "When he could not be the longest and thus to be, and thus to be, the strongest..." (Solomons and Stein)

3. "One does not like to feel different and if one does not like to feel different then one hopes that things will not look different. It is alright for them to seem different but not to be different." (Stein, "Meditations on Being About to Visit My Native Land," 1934)

4. "What a day is today that is what a day it was day before yesterday, what a day!" (Stein, "Broadcast on the Liberation," 1944).

Skinner's theory of the two Steins permitted him to "dismiss one part of Gertrude Stein's writing as a probably ill-advised experiment and to enjoy the other and very great part without puzzlement." The irony seemed to have been lost on our leading reductionist that he was reducing Stein's new style to its leanest, lowest, aridest origin: the knack of automatic writing she had acquired in the course of her undergraduate experiments. Other interpretations might occur to those who believe that behavior—not to speak of literature—might be described in more complex, dynamic ways.

William James thanked Stein for sending him *Three Lives*:

"I promise you that it shall be read *some* time! You see what a swine I am to have pearls cast before him! As a rule reading fiction is as hard to me as trying to hit a target by hurling feathers at it. I need *resistance*, to cerebrate!"

What a challenge to fiction by the brother of Henry James! Hurling feathers, indeed! But, of course, Gertrude Stein had not simply written yet one more work of fiction. She had fabricated a new language, which encountered enough resistance and caused enough cerebration for a gaggle of James'. It could be argued that Stein spent the best part of her professional life opposing the vitalism of James with the mechanistic conception of Loeb. Stein was proud that Marcel Brion had praised her for "exactitude, austerity, absence of light and shade, by refusal of the use of the subconscious."

Her ode to joy on the Liberation: "What a day is today that is what a day it was day before yesterday, what a day!" is not only pure Stein, but is couched in the language of our century: short repetitive sequences. Short repetitive sequences run through modern literature from Morgenstern to Vonnegut, from Beckett to Pinter; they charge the beat of modern music from rock to Philip Glass. Short repetitive sequences also constitute the language of our genes, when we talk DNA or RNA we speak pure Stein. We would not be surprised to hear a molecular biologist explaining a stretch of DNA in the one-letter code of nucleic acids: What a TAA is ATAA that is ATA what a TAA it was, what a TATAA, what a TAA, what a TATAA!

Gertrude Stein worked in the manner Loeb attributed to Diderot: she wrote disinterested sentences the sound of which no false note of personality was permitted to disturb. Her champions praise her as the last daughter of the eighteenth century and the herald of the twentieth (Brinnin, 1959). Like the cubists, she had broken the common plane of thought to make compositions from its basic verbal elements. As Loeb reduced the structure of living things to the "bare molecular world," so Stein reduced language to its bare molecular level where phonemes throb to their own rhythm. Stein's new language expanded the technology of prose as cubism expanded that of painting. Her language was to the painting of her day as a digital display is to the analogue graphic of ours: signs of the present with an eye to the future.

Gertrude Stein tells us that she had learned from William James that: "science is continuously busy with the complete description of something, with ultimately the complete description of everything" (Stein, 1985). But words, sounds, *things* were not descriptions:

"A daffodil is different from a description, a jonquil is different from a description. A narrative is different from a description... A narrative is at present not necessary." (Stein, 1931).

Neither modern writing nor modern art were going to *describe* things, that was the job of science. Nor should the moderns write narrative, the last century had smothered words with stories. No, the task of writers in the twentieth century was to free words and images from the baggage of sentiment. Free from myth, meaning, and station, each jonquil or daffodil, each magpie or pigeon, could stand fresh on the page. A rose is a rose is a rose—and with that line, said Stein, a rose was red for the first time in English poetry for one hundred years. She had re-invented the rose and was free to create an army of roses at will, each red for the first time.

Stein's reduction of language to its basic elements would not have pleased William James. As if in obedience to the rule of parsimony, she had pared from words their moral and aesthetic associations. Did she remember that her Cambridge master had preached to the Unitarians a *summa* against Occam (James, 1974):

"But if the religion of exclusive scientificism should ever succeed in suffocating all other appetites out of a nation's mind...that nation, that race will just as surely go to ruin, and fall a prey to their more richly constituted neighbors, as the beasts of the field, on the whole, have fallen a prey to man.

"I myself have little fear for our Anglo-Saxon race. Its moral, aesthetic, and practical wants form too dense a stubble to be mown down by any scientific Occam's razor that has yet been forged. The knights of the razor will never form among us more than a sect..."

Gertrude Stein was rather richly constituted, herself. Separated from any of her classmates at Radcliffe and Hopkins by barriers of religion, sex, and appetites, she proceeded to outshine them all from an ocean away. With the inner pluck of her Melanctha or black Teresa, she found new ways to clear the dense stubble of Victorian piety.

In 1988, over the main entrance to the Lillie auditorium of the Marine Biological Laboratory stood an enlarged photograph of "the big fat girl" on her collecting trip to Quissett. In that centennial year she shared place of honor with the founders of the MBL and with Jacques Loeb. The men are remembered by official, bronze tablets describing their achievements. If Loeb's tablet reminds us what it is that scientists discover, Stein's picture reminds us that our students frequently discover themselves.

Loeb and the mechanists were no less revolutionary than Stein and the cubists; both movements changed the rules of reason in the name of the twentieth century. In the years before 1914, mechanist and cubist alike marched under the banners of secular humanism to anthems of the machine. Both groups set out to control life, to re-create it, content no longer to remain in tune with its laws. The mechanistic conception of life meant that in the new century science would be in the business of forming life in the dish. Writing to Ernst Mach in 1890, Loeb announced the mechanist's program:

"the idea is now hovering before me that man himself can act as creator even in living nature, forming it eventually according to his will."

We would call that enterprise "biotechnology." A century later, products of our biological revolution—the army of molecules that James feared—flow in growing numbers from the lab bench to the bedside. Each molecule is like every other, each made according to will, each dictated by the language of the genes and written in short, repetitive sequences: TAA, now TATAA, now now TAAATAAT, now TAAT!

Stein had chimed in half a century ago:

"The characteristic thing of the twentieth century was the idea of production in a series, that one thing should be like every other thing, and that it should be alike and quantities of them." (Stein, 1970)

From drugs to genes, from soup to nuts, our century has made things like every other thing and quantities of them. The writer to whom a rose is a rose is a rose would have appreciated a machine to which DNA is DNA is DNA, or for that matter, a painter who tells us that a can is a can is a can. Wendy Steiner (1978) has called attention to an interview with Andy Warhol in which he told his interviewer that "I think everybody should be a machine. I think everybody should be like everybody...because you do the same thing every time. You do it over and over again."

As Stein and Loeb foresaw, the twentieth century has made it possible to create living things at will, each cell like every other cell and quantities of them. We can do it over and over again; we call the process "tissue culture" and its study "cell biology." The students who collect creatures at Quissett today are cell biologists who can read from ctenophore and human cells the same short palindromes of DNA.

The victories of mechanist and cubist might be considered a mixed blessing. In science, our mastery of animate nature has given us tools with which to solve the cypher of genes, the floor-plan of cells, and the wiring of our nerves. We have just begun to treat cancer, to stop heart attacks, and to cool terrors of the mind. As for the arts, we have lived in the century of Picasso, Stravinsky, and Brecht. But those of the reductionist persuasion cannot escape the nastier fallout of our revolution. Our culture cannot be said to have grown uniformly richer after the coming of Stein and the cubists; we now have Mamet instead of Ibsen and soup cans instead of Renoir. Our writers have so pared serious literature that it is read by very few; our composers have so trimmed their measures that a generation is deaf from tribal chants; we are so used to having machines tell us stories that our most popular tale is told by a cartoon rabbit.

Our reductionist science is in the dock for less trivial reasons; we have undergone what William James called "the great initiation" into the game of guilt. We cannot escape blame for the fouling of our environment, the anomie of our cities, the commercialization of research, and the pedantry of learning. More darkly, still, we remember each August on Hiroshima day the most destructive outcome of a mechanistic physics. Adherents of a mechanistic biology have yet to devise a morality that can protect us from that sort of mischief.

But remembering Stein the student and Loeb the teacher of the MBL ten decades ago, one is struck by the generous impulses that energized their work. The aims of that mechanistic conception were the control of life in order to understand it and to cure its ills, the reduction of life's blooming, buzzing confusion to the manageable laws of chemistry and physics, and the divorce of biology from those doctrines of vitalism and theism that had been used to justify rank and bigotry. Indeed, the tribute Stein paid to Picasso could be paid her and Loeb as well:

"...a creator is contemporary, he understands what is contemporary when the contemporaries do not yet know it, but he is contemporary and as the twentieth century is a century which sees the earth as no one has ever seen it, the earth has a splendor that it never had, and as everything destroys itself in the twentieth century and nothing continues, so then the twentieth century has a splendor which is its own and [they are] of this century, [they have] that strange quality of an earth that one has never seen and of things destroyed." (Stein, 1938)

Literature Cited

Brinnin, J. M. 1959. *The Third Rose.* Addison and Wesley, Boston, MA.

Curtis, W. C. "Good Old Summer Times at the MBL," *The Falmouth Enterprise*, Falmouth, MA: August 12, 1955.

Freud, S., and J. Breuer. 1966. Studies in Hysteria. Reprint of the 1895 edition in *The Standard Edition of the Complete Psychological Works*, J. Strachey and A. Freud, eds. Hogarth Press, London.

James, W. 1974. *Essays on Faith, Ethics and Morals.* New American Library, New York.

Lewis, S. 1924. *Arrowsmith.* Harcourt Brace, New York.

Loeb, J. 1916. *The Organism as a Whole: From a Physico-Chemical Viewpoint.* G. Putnam and Sons, New York.

Loeb, J. 1965. *The Mechanistic Conception of Life*, reprint, Belknap, Cambridge Press. Press of the Harvard University.

McGrath, W. J. "Peter Gay: Freud," *New York Review of Books*, New York: August 12, 1988.

Pauly, J. P. 1987. *Controlling Life: Jacques Loeb and the Engineering Ideal in Biology.* Oxford University, Oxford.

Skinner, B. F. 1935. "Has Gertrude Stein a Secret?" *The Atlantic,* Boston: April, 1935.

Solomons, L. M., and G. Stein. 1896. Normal motor autonomism, *Psychol. Rev.* (Harvard Psychological Laboratory) **2**: 492-512.

Stein, G. 1931. *How to Write.* Plain Editions, Paris.

Stein, G. 1933. *The Autobiography of Alice B. Toklas.* Harcourt Brace, New York.

Stein, G. 1938. *Picasso*. B. T. Batsford, London.

Stein, G. 1970. *Paris: France*. Liverright, New York.

Stein, G. 1972. *Selected Writings*. Vintage Books, New York.

Stein, G. 1985. "Lectures in America," (introduction by W. Steiner), reprint, Beacon Press, Boston. Originally published by Random House, New York.

Steiner, W. 1978. *Exact Resemblance to Exact Resemblance: The Literary Portraiture of Gertrude Stein*. Yale University Press, New Haven, CT.

Thomas Hunt Morgan
(1866 - 1945)

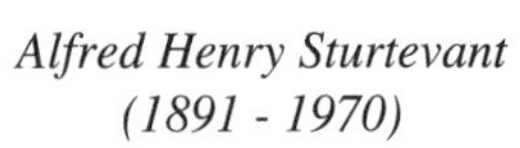

Alfred Henry Sturtevant
(1891 - 1970)

Genetic Maps—Fruit Flies, People, Bacteria, and Molecules: A Tribute to Morgan and Sturtevant

Introduction by Harlyn O. Halvorson

When we celebrated the centennial of the Marine Biological Laboratory, I recalled that the Laboratory arose only a few decades after Charles Darwin called attention to evolution through his book, *Origin of the Species*, and Gregor Mendel described the principles of heredity. Darwin's book received considerable publicity; Mendel's principles escaped the attention of the scientific world. An emerging biological field, soon to be called genetics, attracted the interest of early members of the MBL community, as did the ideas of evolution and experimental embryology. While these ideas remained separate in the academic environments of universities, they merged comfortably at the MBL.

Thomas Hunt Morgan was one of the most gifted young student investigators to come to the MBL that first summer of 1888. He was born in 1866, the same year that Mendel published his first studies of pea inheritance.

Morgan became one of the early leaders in the burgeoning field of experimental embryology. He did many now classic experiments to elucidate the mechanisms of early embryonic development and to describe the natural regeneration of animal parts. In so doing, he studied an almost unbelievable variety of animal species and helped to popularize many organisms that have now become firmly established as marine models.

Although trained as a zoologist, Morgan was known for his regeneration theories and embryology work, and, primarily, for his contributions to Mendelian genetics. Much of his early research was carried out at the MBL, where he came each summer from 1890 until his death in 1945. He became an MBL Trustee in 1897, and remained so involved until his death.

In mid-career, Morgan turned to genetics to seek explanations of evolution and development. He was one of the first scientists to understand the role of a model system. He introduced the fruit fly, *Drosophila*, as a main model of modern genetics. Modern analytic genetics was essentially invented by Morgan and his co-workers in the "Fly Room" at Columbia University and later at Caltech. Morgan himself never ceased to view the mechanisms of heredity and genetics as the means to understanding the more complex events of embryonic development. Later in his career, when he received the Nobel Prize in physiology in 1933 for his theory of the discrete gene as the hereditary and basic functional unit of the cell, he expressed the idea that differentially segregated cytoplasmic materials of the early dividing embryonic cells selectively and sequentially activate certain genes in development. To this day, his wider theory remains a central paradigm of all current research in developmental biology.

Morgan "discovered" **Alfred Henry Sturtevant** while the teenager was attending a general zoology class at Columbia University. Sturtevant had intrigued Morgan (who was not fond of teaching but had temporarily taken over the class for a colleague) with his observations of horse breeding patterns on his family's farm.

In 1910, Sturtevant and C. Bridges were given desks in Morgan's laboratory. It was in this so-called "Fly Room" at Columbia where the three worked for the next 18 years. Sturtevant described the room during a lecture on Morgan at the MBL in 1966: "this was a room 16x23 feet in which there were eight desks. There was a place where we cooked fly food, and there were usually at least five people working in there. Bridges and I practically lived in this room; we slept and ate outside, but that was all. And we talked and talked and we argued, most of the time. I've often wondered since, how any work at all got done with the amount of talking that went on, but things popped."

The group was soon joined by H. J. Muller. In this exciting atmosphere, Sturtevant offered the first explanation of the phenomenon of multiple allelism, ascribing it to alternative states of the same gene. He continued to work with Morgan for many years in the "Fly Room" and made important advances in sex-linkage and mutation. Sturtevant is credited with constructing the first chromosome map in 1913.

Sturtevant studied *Drosophila* during his entire scientific career, which spanned more than 50 years. But he also studied horses, snails, rabbits, moths, flowers, and humans.

The result of the quantitative studies of Morgan and Sturtevant was a map of the four chromosomes of the fruit fly which, by the mid-1940s was already well developed.

The stage was now set for the next revolution, which began with the work of Joshua Lederberg. Introduced to bacterial genetics as an undergraduate by Francis Ryan, Joshua Lederberg went on, in a landmark Ph.D. thesis with Ed Tatum, to discover genetic exchange between certain strains of bacteria—a process of sexual exchange called by Luria in 1947 to be "among the most fundamental advances in the whole history of bacteriological science." Only a few years later, at the University of Wisconsin, Lederberg showed, with his student Norton Zinder, that genetic information and recombination could occur between bacteria when a bacterial virus transmitted a part of the bacterial chromosome from one strain of bacteria to another. In only a short period of time, two of the major mechanisms of genetic exchange in bacteria—conjugation and transduction (meaning leading across)—had been described. A decade of exciting activity in Dr. Lederberg's laboratory and in others set the stage for the current revolution in molecular genetics and for the investigation of the chemical nature of genetic material, the nature of mutation and recombination, and the regulation of gene activity.

Joshua Lederberg received his B.A. from Columbia University (1944) and his Ph.D. from Yale University (1947). From 1947 to 1959, Lederberg was professor of genetics at the University of Wisconsin. In 1959 Lederberg moved to Stanford University Medical School and became chairman of the Department of Genetics. There he collaborated with E. A. Feigenbaum on experiments in artificial intelligence, pioneering the development of "expert systems." From 1978 - 1990 he was president of The Rockefeller University, where he oversaw research and graduate education. Lederberg was awarded the 1958 Nobel Prize for his studies on genetic recombination in bacteria. He has received numerous honorary degrees and has been elected to membership in many scientific societies, including the National Academy of Sciences and the American Philosophical Society. He has worked on many government advisory committees dealing with health research and science policy, and has worked in a similar capacity with the World Health Organization.

Genetic Maps—Fruit Flies, People, Bacteria, and Molecules: A Tribute to Morgan and Sturtevant

Joshua Lederberg

The Rockefeller University

THE MBL CENTENNIAL CELEBRATION was a wonderful reunion, at every imaginable level. My visits to Woods Hole began forty-five years ago; but I always felt like a little bit of an interloper. I never found a very good excuse to work on sea urchins or the giant squid or any of the other favorite organisms of this place. It was bad judgment to pick bacteria, which provide very little rationale for the enjoyment of the MBL: but I would visit the laboratory every now and then regardless. Unfortunately there are still very few organisms that live in the sea whose genetics have been developed. It's time we change that.

I'm going to discuss genetic maps, and I assume that the reader has a general knowledge of genes and chromosomes, but does not know so much of the details of DNA chemistry that my simplifications in that realm will be irritating. Just relax, and read a story of genetic maps, and how our understanding of the maps has developed. This essay is based on an evening lecture, not a semester course in molecular biology (see bibliographic note).

Thomas Hunt Morgan came to a chair in Zoology at Columbia University in 1904 at the behest of the renowned cytologist, E. B. Wilson. Shortly thereafter he began breeding *Drosophila* with the intention to make it the exemplar of genetic investigation. And he set up the famous fly room, shown in Figure 1. This picture was purportedly taken (according to Curt Stern's notes) secretly, without Morgan's knowledge. He was supposed to have been somewhat

Figure 1. *T. H. Morgan in the "Fly Room" at Columbia University. (Reproduced with permission from* Am. Zool. ***26****: 573-581, 1986.)*

phobic about the matter. The picture was taken with a camera that was concealed in one of the fly incubators.

As you can see, Morgan is almost concealed by the cream bottles that were used for growing flies at that time. It is a small exercise in personal nostalgia to recall where the fly room was. It was on the sixth floor of Schermerhorn Hall, Room 613, close to where I started my own research as a college student in 1941. Between 1908 and 1928, this room was the creche of the chromosome theory of heredity. I've started a campaign at Columbia that there should be some commemoration of this remarkable place—the Thomas Hunt Morgan fly room.

In 1911—30 years before my arrival—Alfred Sturtevant was also an undergraduate at Columbia. He had joined Morgan's laboratory together with Calvin Bridges and Herman Muller. These four people were really the steam engine for the origination of our modern theory of genetics. Here is Alfred's own statement of how he came upon the genetic map:

THE "FLY ROOM"

In 1909 Castle published diagrams to show the interrelations of genes affecting the color of rabbits. It seems possible now that these diagrams were intended to represent developmental interactions, but they were taken (at Columbia) as an attempt to show the spatial relations in the nucleus. In the latter part of 1911, in conversation with Morgan about this attempt—which we agreed had nothing in its favor—I suddenly realized that the variations in strength of linkage, already attributed by Morgan to differences in the spatial separation of the genes, offered the possibility of determining sequences in the linear dimension of a chromosome. I went home and spent most of the night (to the neglect of my undergraduate homework) in producing the first chromosome map, which included the sex-linked genes y, w, v, m, and r, in the order and approximately the relative spacing that they still appear on the standard maps (Sturtevant, 1913).

—Quoted from Sturtevant (1965)

Linkage is the consequence of genes being located on the same chromosome not far from one another. Such genes will co-segregate in crosses unless some other event happens. That other event is called "crossing-over," and the frequency of crossing over is a function of the distance between genes on the chromosome, and that is translated to distance on a genetic map. As Sturtevant notes above, his first maps show a number of genes located in the order and approximately the relative spacing that they still appear on a standard map (Figs. 2, 3).

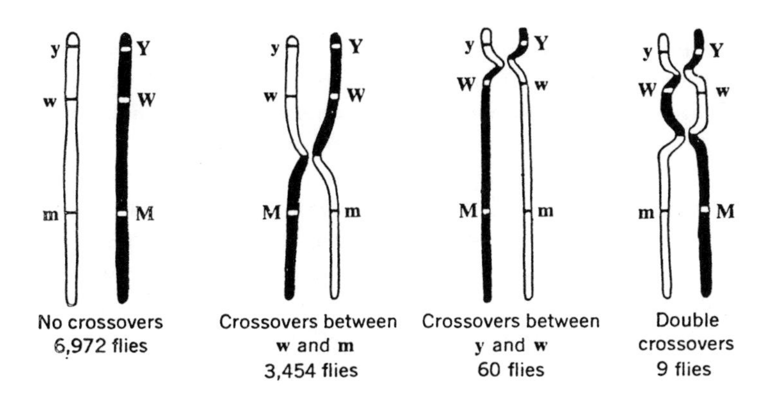

Figures 2 and 3. *Examples of genetic maps. (Reproduced with permission from* Am. Zool. ***26****: 573-581, 1986.)*

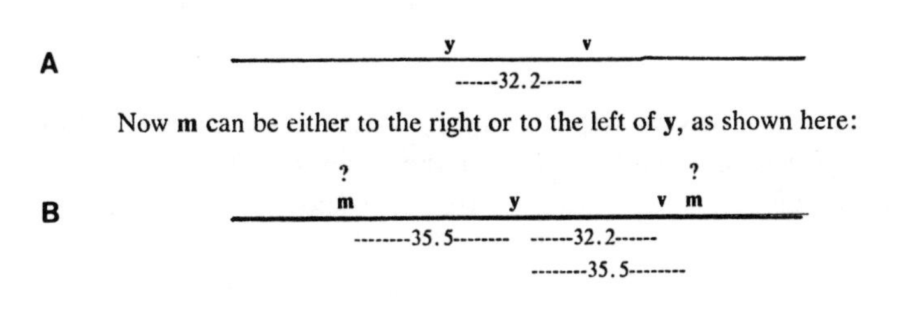

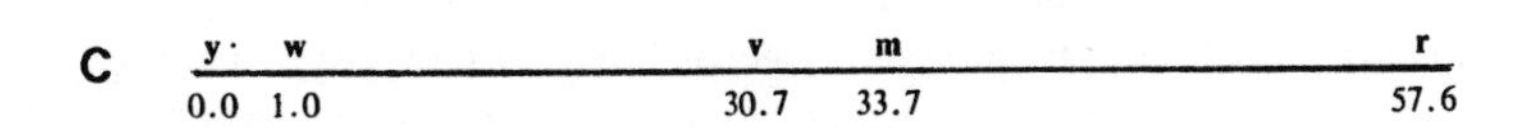

If markers are perfectly linked they will appear at the same point on the chromosome. If they separate from one another 1% of the time, we say they are one linkage unit, one centimorgan, apart. To verify the linear order of markers, one has to take markers not just by twos, but at least by threes to resolve ambiguities as to the specific sequence. When they do all fit together reasonably well by a series of three point tests and higher, we have a corroboration of the linear map.

At that time *Drosophila* was known to have four pairs of chromosomes. A peculiarly looking chromosome called the Y was present in the male and not the female, and a few sex-linked markers, that is to say genes, were located on the X chromosome. E. B. Wilson had first hypothesized that the inheritance of color blindness could be explained by the position of that marker in the sex chromosome of the human. The sex-linked markers, besides bolstering the chromosome theory, are technically much easier to study; results can be obtained in one generation rather than two generations of breeding.

By 1915, we had the listing of the linkage groups of *Drosophila*, as shown in Figure 4. The chromosome picture hadn't changed, but now there are 85 genes. The genes fell into four linkage groups, a rational finding because four chromosome pairs were available. Only a couple of genes could be identified with the tiny chromosome number 4. One thing that is evident immediately is that the locations of the genes for factors involving different developmental outcomes make no particular sense whatsoever. Many of the chromosomes had genes for color; every one of them had genes for wing shape, and so on. Again there seems to be no developmental sense to the map. We know today that if you compare different species of *Drosophila* you will find that the genetic locations of homologous genes are terribly scrambled. You can look for correspondence between any one of the chromosomes in one species and another. You will find that, typically, the most obvious genetic difference between species is the structural change in the chromosomes where pieces of chromosomes have been moved around to many different places. So for many years the idea that the comprehensive genetic map could enhance our insight into the developmental role of the genes was substantially frustrated. Fine structure studies have been more informative.

As Harlyn Halvorson mentioned in his introduction, gene mapping continued well through the '40s, having been facilitated by Theophilus Painter's serendipitous discovery, in 1933, of the giant chromosomes of the salivary glands of *Drosophila* (Fig. 5).

GROUP I

Name	Region Affected
Abnormal	Abdomen
Bar	Eye
Bifid	Venation
Bow	Wing
Cherry	Eye color
Chrome	Body color
Cleft	Venation
Club	Wing
Depressed	Wing
Dotted	Thorax
Eosin	Eye color
Facet	Ommatidia
Forked	Spines
Furrowed	Eye
Fused	Venation
Green	Body color
Jaunty	Wing
Lemon	Body color
Lethals, 13	Die
Miniature	Wing
Notch	Venation
Reduplicated	Eye color
Ruby	Legs
Rudimentary	Wings
Sable	Body color
Shifted	Venation
Short	Wing
Skee	Wing
Spoon	Wing
Spot	Body color
Tan	Antenna
Truncate	Wing
Vermilion	Eye color
White	Eye color
Yellow	Body color

GROUP II

Name	Region Affected
Antlered	Wing
Apterous	Wing
Arc	Wing
Balloon	Venation
Black	Body color
Blistered	Wing
Comma	Thorax mark
Confluent	Venation
Cream II	Eye color
Curved	Wing
Dachs	Legs
Extra vein	Venation
Fringed	Wing
Jaunty	Wing
Limited	Abdominal band
Little crossover	II chromosome
Morula	Ommatidia
Olive	Body color
Plexus	Venation
Purple	Eye color
Speck	Thorax mark
Strap	Wing
Streak	Pattern
Trefoil	Pattern
Truncate	Wing
Vestigial	Wing

GROUP III

Name	Region Affected
Band	Pattern
Beaded	Wing
Cream III	Eye color
Deformed	Eye
Dwarf	Size of body
Ebony	Body color
Giant	Size of body
Kidney	Eye
Low crossing over	III chromosome
Maroon	Eye color
Peach	Eye color
Pink	Eye color
Rough	Eye
Safranin	Eye color
Sepia	Eye color
Sooty	Body color
Spineless	Spines
Spread	Wing
Trident	Pattern
Truncate intensf.	Wing
Whitehead	Pattern
White ocelli	Simple eye

GROUP IV

Name	Region Affected
Bent	Wing
Eyeless	Eye

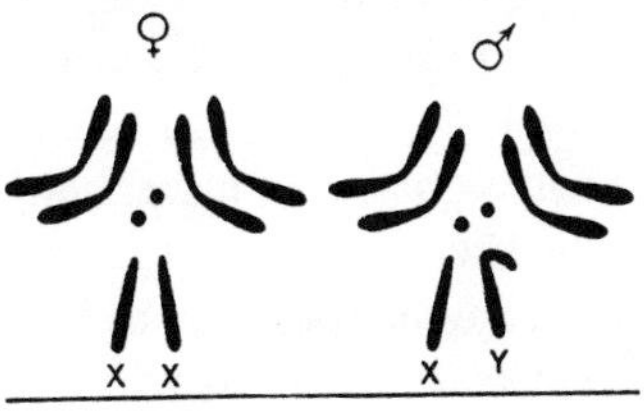

♀ ♂

X X X Y

Figure 4. Drosophila *linkage groups. (Reproduced with permission from* Am. Zool. ***26****: 573-581, 1986.)*

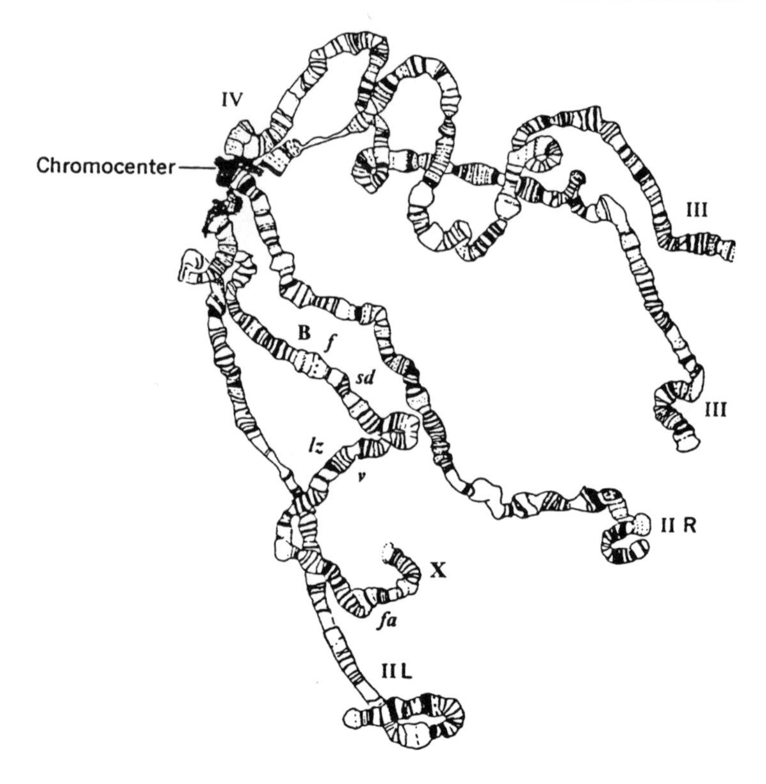

Figure 5. *Giant chromosomes of the salivary glands of* Drosophila. *(Reproduced with permission from* Am. Zool. ***26****: 573-581, 1986.)*

Drosophila was developed as a standard or type organism for experimental investigation in genetics because of convenience in handling. It was of a size such that you could just about see it with the naked eye, and a really clever and astute observer could pick out mutants by inspection. You didn't have to provide enormous amounts of food or other logistical care to sustain thousands—or tens of thousands—of individuals. Before the introduction of bacteria into genetics, those were regarded as astronomical numbers of organisms for genetical research. But on top of that, it now had one more treasure to offer, namely the giant chromosomes that just pop out for no reason other than that God was kind to us during the cytogenetic research on the salivary glands of the dipteran larva. With the discovery of these large chromosomes, it suddenly became possible to study the anatomy of the chromosome in finer detail, to at least two orders of magnitude, even just using the light microscope.

Drosophila scientists were thrilled by Painter's discovery. It was so exciting that Morgan, having just received the announcement that he was to be the recipient of the Nobel Prize in physiology or medicine, decided to defer going to Stockholm because he didn't want to lose out on the opportunity to act very, very quickly, exploiting Painter's finding. He wanted to include the salivary gland chromosome mapping in the Nobel address that he gave the following summer in 1934.

The experts found that they could identify even small segments of chromosomes by their specific banding pattern. If a chromosome was broken and a piece translocated to another place, they could track that rather well. They could even eventually locate individual genes to individual bands. I guess that there were about a thousand bands all together so there is that degree of subdivision of the genome that becomes available by this technique of visualizing banding patterns.

Figure 6 shows photomicrographs of one of the salivary chromosomes. Another attribute, again God-given, is that the chromosome pairs in the salivary glands are fused and aligned in much the same fashion as happens transiently during meiosis. So if you have structural changes whereby part of one chromosome has been broken off and translocated to another one, that is faithfully replicated and magnified in the salivary chromosome. When the altered giant chromosome participates in somatic pairing with the other normal chromosome of the diploid pair, you get conformations like that shown in Figure 6, which indicates that in one of the parents this chromosome has been broken at that point. So you can really track quite small changes in the movement of the chromosome parts from one to another of that set.

The giant salivary gland chromosomes deferred a latent crisis in the development of genetics: that the usefulness of *Drosophila* as a tool, had nearly been exhausted because of the methodology available at that time.

Nevertheless, scientists were stymied about going still further into the fine chemical structure of the gene in *Drosophila.* They lacked a means of subdividing the chromosome that was sufficiently detailed so that you could look at individual genes or look into their structure by either the cytological or genetic methods available.

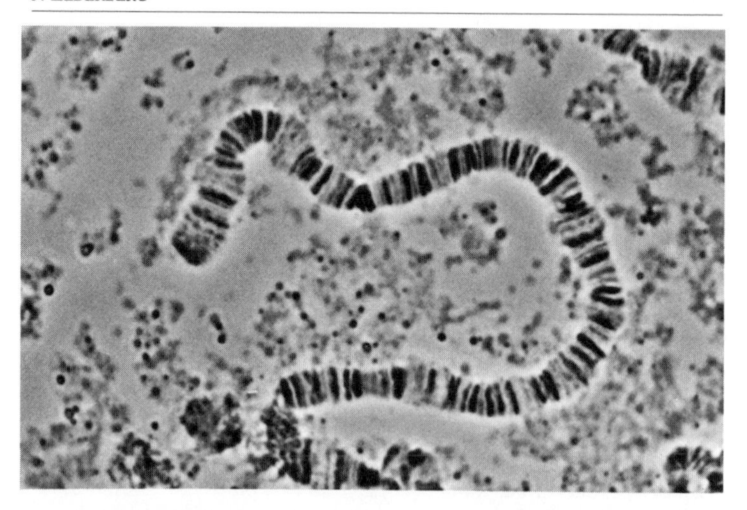

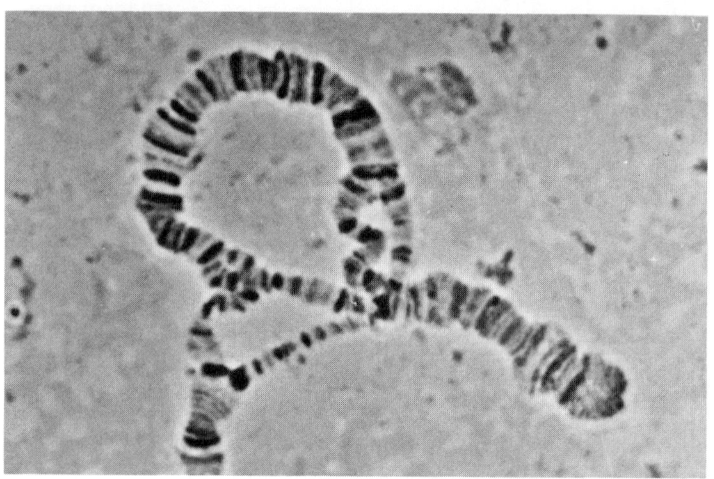

Figure 6. *A photomicrograph of one of the salivary chromosomes of* Drosophila. *(Reproduced with permission from* Am. Zool. ***26****: 573-581, 1986.)*

New opportunities arose by turning to other "organisms" from the world of microbiology. The classic work "PATOOMB" (Cairns *et al.*, 1966) documents the historic development of bacteriophage as a system of biological inquiry, largely through the impetus of Max Delbrück.

In 1946 at the July symposium at Cold Spring Harbor, both Delbrück, and (essentially independently) Al Hershey reported on a new phenomenon that had cropped up in their investigation of the T family of phages. The T-even phages, T2, T4, and T6, arbitrarily plucked out of the sewers of the world, turned out to be, in fact, rather closely related to one another in their genetic behavior; they are more or less interchangeable. It turns out that if you start with two virus strains in T4, one that had achieved a mutation that changed its host range and another that had achieved an alteration that changed its plaque morphology (one of the so-called R mutants), and infect a single bacterial cell with at least one each of the two parental types, you would find new combinations—genetic recombinant phages—in the crop of viruses that was generated from that cell.

In the population emerging from the mixed infection, you would get some wild type from putting the two mutants together, and you would get some that were double mutant—a virus that was altered both in host range and in its plaque morphology. When this work first appeared there was some skepticism, especially from Delbrück, about a recombinational interpretation. He had been investigating interference between phages in such depth that he viewed the results from that standpoint. He had discovered another phenomenon in which it appeared that once a single phage had penetrated a bacterium, it would tend to exclude other phages participating in growth on the same bug. So his first hypothesis to interpret the mixed infection results was not genetic recombination at all, but some effect of one virus particle, from outside the cell, on the growth of the virus of the other genetic type.

But Hershey followed his work up in some detail, and a few years later he had built a genetic map. His investigation of a variety of host range and R mutants provided the basis for the recombinational genetics that we have seen ever since. Although the T4 genetic map is described in circular terms, the actual DNA of any given particle of T4 as you see it wrapped up within its own envelope, is, in fact, a linear piece of DNA, but it is cut as a permutation from a circular map of that organism. When the DNA replicates it simply rolls over and over again from its starting point, so you have pieces cut from a sort of rolling circle of spaghetti to wrap them up into a single bacteriophage particle. This replication proceeds continuously, resulting in a hundred or so progeny from each infected bacterial cell.

In T4, the individual genetic cross is hard to control: when you put two virus particles in, you get a few hundred progeny out, and only with difficult tricks can you snapshot the intervening process. About five successive rounds of mating occur in the pool of virus genomes that develop within the mixedly infected bacterium. I'm not going to go into the details of bacteriophage replication and recombination here, but I do want to describe the way in which Seymour Benzer, starting in about 1955, exploited the technical opportunities T4 recombination afforded "to run the genetic map into the ground."

Benzer noted a certain class of R mutations, which emphasized the laborious task of looking at individual plaques in order to type them. Instead, he found a method of selection. Certain hosts, in particular *E. coli* K-12 (a different bacterial strain than was being used for the original growth of the phages), would not sustain the growth of any of the RII mutants, but would accept the wild type. Therefore, he counted what fraction of the phage particles would give plaques on this differential host, K-12, compared to what they give on the standard host, B. He could selectively choose rare wild type phages produced as crossovers or recombinants between different RII mutants, even if the mutational sites were very close together and the frequency of recombination low.

His method was so precise that he could easily pick up recombination when it occurred at distances of only .0001%. He collected hundreds of these RII mutants, testing them against one another. From his library of hundreds of mutants, he could find mutants that behaved as if they were adjacent ultimate units of the genetic map, defining a quantum of recombination. If you took two mutants and they were very close together: either they didn't cross over at all, to a sensitivity of .0001%, or they crossed over with a frequency of at least .01%. So it appeared as if there were a certain grain in the structure of the genetic material, namely, mutations that corresponded to changes in adjacent nucleotides of the DNA sequence. You can't have two markers closer together than being adjacent nucleotides—unless they are identical. This conclusion still stands, although we must now take into account gene conversion and other complications.

Besides many stretches of closely adjacent mutants, Benzer found others that seem to cover overlapping segments. These mutants were deletions, and they proved to be an enormous technical help, and also corroborated the linearity of the sequence. Deletion mutants would fail to give recombinants with each of the single mutants within the genetic region spanned by the deletion. Deletions were invaluable in the actual construction of a genetic map of the RII gene of phage T4. If you have libraries of several hundreds of these mutants, you really don't want to have to cross every one with every other one to map them. It is enough to see whether a given mutant fails to recombine with a certain deletion to know whether it falls within this segment or outside of it.

This work was the first to show mutational events that corresponded to the physical model of the chromosome that was now being developed in terms of the DNA structure.

The total DNA sequence has now been pretty well worked out. All the mutants that Benzer worked on were nucleotides within the RII segment. The map as a whole, however, is some 165 times larger than that one segment; there are 166,000 in all of the base pairs and almost all of the genes of phage T4 have been located and identified in functional terms. Therefore, mutations having some impact in the development of the organism had been localized on the T4 map. They can be related in their connection with one another: for instance, what a wonderful thing it is that we have a complex of genes located together that all have to do with the construction of the head. As you go down the map you do begin to see locally, regionally, that genes of related function are, in fact, often physically coordinated in the structure of the chromosome. But you need a very high resolution to be able to see these relationships and there are exceptions to it. For some of the exceptions you find that the gene seems way out of place, but nevertheless can interact developmentally with something on the other side of the map, thus preserving a connection between gene location and function.

Because phage T7 is smaller than T4, it has been examined in somewhat more detail: it is only 39,936 base pairs long. It has also been run into the ground both chemically and genetically. In general terms, one can classify the characteristic genes of the organism as falling into an early, a middle, and a late class depending on when the function of those genes have to be elaborated in terms of the needs of the initial steps of infection, replication, the transcription of the phage

DNA, and finally the components needed for the assembly of the bacteriophage itself in the maturation of the phage particle.

I now turn to another system of mapping that Harlyn also referred to briefly in his introduction. I was working with Ed Tatum at Yale. In 1946, at the very same meeting at Cold Spring Harbor where Delbrück and Hershey reported on recombination phage in T4, I had the exciting opportunity of talking about our work on recombination in the bacterium *E. coli* K-12. In fact, Ed and I probably wouldn't have published a work quite that quickly if it weren't for the excitement that was being generated by the Delbrück and Hershey reports. Our report was not preplanned on the program of the meeting. I had done my first really successful experiment on crossing *E. coli* strains at about the end of May, then during all of June it had been repeated at least a dozen times. There was no doubt about the reality of the observations that *E. coli* bacteria underwent genetic recombination, but we still felt a little bit edgy about rushing to print after only six weeks of an inquiry.

But the data were there. They were quite solid and the mood was right, so Ed and I agreed that it was perfectly appropriate to publish. That really was quite an exciting summer. I devoted almost the entire next year to developing a genetic map in *E.coli* K-12, using essentially the same paradigm that Sturtevant had used in 1911. I wrote up my findings at the MBL Library during the summer of 1947; that was my doctoral thesis at Yale. The map of *E. coli* genes that we published in *Genetics* is a modest subset, but it can be superimposed on today's map.

We continued the mapping work by introducing some further markers; but by 1951 we were facing a disaster. The original markers that I indicated previously had mapped linearly; but when we had a few others like maltose, streptomycin resistance, xylose, and mannitol fermentation, they made no sense at all. In the *Genetics* paper I represented it as a branched map, but unfortunately, some people didn't read the figure legends carefully. As we insisted at the time, this diagram is purely a formal represention of recombination frequencies, and does not imply a physically branched chromosome. What it does imply is that we were in trouble. That is to say that anomalies in recombination frequencies appear in the way that the consequences of crossing were not working out simply. There was something not understood. My own mental model of conjugation in 1946 was borrowed from paramecium: the complete transfer of a whole nucleus from one cell to another. But Jacob and Wollman showed that there is a progressive transfer of that chromosome, starting from a particular point of origin. It takes 100 minutes for the full consummation of this transfer.

The French refer to "coitus interruptus," which is how one studies the kinetics of gene transfer in bacteria. If you simply give the culture tubes in which the mating is occurring a violent shake at a given time, you physically disrupt the mating process. Then only as much DNA as has already penetrated the female cell can further participate in recombinational processes. An additional detail: a single strand of the double strand helix of the male DNA is replicated as it enters the female cell.

I now turn to today's comprehensive map of the *E. coli* genome, the compilation of which we owe so much to Barbara Bachmann.

First of all, mapping *E. coli* is done today by taking full advantage of the temporally sequenced transfer of markers. The position on the linkage map is calibrated by how many minutes it takes for the first appearance of a marker donated by the male cell to appear among the recombinants (these stretches of map are in groups of 10 min). Each minute equals about 50,000 base pairs; *i.e.*, roughly 1000/second.

The process is sufficiently fragile that hardly any crosses really go to completion to the full 100 minutes of the map length. And so you can see that there were certain complications about what was going on in conjugation in *E. coli* that confound the details of mapping by Sturtevant's method, but now almost everything has fallen into place. In bacterial conjugation we have found a different principle of mapping that we can correlate with the method of mapping by the frequency of crossing over between two markers.

Mapping can be difficult if you have a marker in a genetic region that is not otherwise well populated with genes. You can still find out where that particular gene is by using the temporal sequence method. That tells you approximately where the gene is and what to look for in terms of very closely linked markers if you want to then establish within a millimorgan exactly where it is on the map. This is a fairly arcane presentation, and you might ask, what is the use of it? It is interesting to know that we have gotten well over a thousand different genes mapped in *E. coli.* We are reaching the point where we can expect to saturate the map. This information has ended up having a surprisingly practical significance. As arcane as it seems, you can't do serious biotechnology using the recombinant DNA paradigm without exactly these kinds of maps that define where the markers are into which you are going to splice new factors, finding the promoters to genes you want expressed, and constructing specific recombinants. So it has ended up being far from an abstruse, arcane, and useless outcome, whatever golden fleece awards might have been conferred on studies of the sex life of bacteria in the 1950s.

Another mode of genetic recombination, transduction, was first discovered by Norton Zinder in my laboratory (Zinder and Lederberg, 1952). It was his doctoral thesis at the University of Wisconsin. We found it while looking for another organism that could be crossed like *E. coli* could be crossed, and we wanted some taxonomic diversification. It seemed important to look at an organism that had a certain medical significance, whose serology was very important, and had been very well worked out as in the case of *Salmonella*. So there were both a number of practical aspects, and a certain background of microbiological knowledge, that would really make one ache to cross *Salmonella* as well as *E. coli*. We used the same paradigm as was used in *E. coli* K-12, and it just didn't work at all! It took a year or two of floundering around to realize that they were not crossing the same way *E. coli* K-12 was. Instead, there was another phenomenon.

A bacterial virus was living in a lysogenic state hidden within the chromosome of one of the strains of *Salmonella*. This phage would occasionally be liberated by the medium, would grow on the other *Salmonella* strain, then that phage crop grown on that other *Salmonella* strain would pick up some of its genes and then move them back to cells of the first strain—and give rise to a number of genetic recombinants by that route. A single phage particle can package about 1% of the total genome in the *Salmonella* system. In *E. coli* another transduction system of similar value was exploited by Yanofsky and Lennox (1959). In *E. coli* you could now relate transduction recombination data to the crossing maps that I mentioned before, but you now had the advantage of much higher resolution because only 1% of the genome is in any given phage particle. Two markers are only cotransduced—they only appear to be linked—when they are very, very close together. And you can then study the detail of crossing over that then occurs within that 1%-interval of the amount of DNA that is, in fact, carried within a given phage particle.

So Yanofsky and Lennox used that system for an enterprise rather like what Benzer had done previously in T4, down to almost the same level of resolution. Yanofsky in particular was able to find mutons that represented mutations in adjacent or nearly adjacent nucleotides.

By studying the tryptophan synthetase gene and getting thousands of mutants within that particular gene, and crossing them by transduction, he was able to relate changes in the genes according to their map position with changes in the amino acids in different positions of the protein itself. So he confirmed the linearity of the genetic map, and the colinearity of the genetic map with the amino acid sequence of the protein: there was clearly a one-to-one relationship. That is to say the information in the DNA was not only encoding for the general properties of the protein that it was responsible for, but it was determining nucleotide by nucleotide which amino acid was going to be present at a given point in the primary structure of the protein.

This set the stage and provided the biological background needed for the cracking of the genetic code, which proceeded in the early 1960s in a rather more direct chemical examination by Nirenberg and others. During my lecture I showed some examples of map structure having to do with physiological consequence. In the map, *trp* A, B, C, D, E referred to five different enzymes involved in tryptophan biosynthesis. The genes for the enzymes appear stacked one next to another on the bacterial chromosome. We get some inkling about why that should be, not so much as to how it came about (we weren't there at the creation), but what sustains it by the understanding that we are beginning to have about the control mechanisms for the expression of this gene sequence.

E. coli is very clever. It can synthesize its own tryptophan by turning on the genes for these enzymes, and it does so in minimal media—media that are not already loaded with that amino acid. But *E. coli* shows great economy in the deployment of its synthetic resources. When tryptophan is available in the outside medium, it shuts off its own internal production. It saves the need to synthesize that whole set of enzymes that is relevant to making that particular growth factor and nothing else.

There is a promoter: the DNA sequence that represents the point of binding of the RNA polymerase, the enzyme that is involved in the transcription of the DNA information into RNA. There is also an operator: a point on the chromosome that is sensitive to amino acids in the environment and is able to exert a negative influence on the further transcription of the DNA if there is trytophan bound to a repressor protein that then binds to the operator site. In addition, there is an attenuation mechanism, which at the translation stage is also sensitive to the level of tryptophan in the environment and shuts off translation if tryptophan is present, or, more accurately, allows it to read through a stop point if tryptophan is absent from the environment.

Thus the co-location of these *trp* genes in a related biochemical sequence is connected to their co-adaptivity to the organism. It is easy to see why the organism should want to turn these enzymes on and off *en bloc* rather than have different mechanisms to deal with them separately. One way to do that is to have them co-located on the DNA sequence. The genomes of both viruses and prokaryotes contain innumerable examples of these kinds of complexes.

I'm going to go now to human chromosomes, and I'm going to jump very quickly, although the pace of events was at first inordinately delayed and then very rapid once the plug was pulled. It was not until 1956 that Tjio and Levan published the correct chromosome number in man: 2n=46. That's a staggering statement! Many of you were born by then, so this event occurred within your own memory. We were so muddled about how to look at the nucleus of the human cell, that we carried around a model that had the wrong number. What could be worse than that to confuse human biology!

However, when 23 pairs were identified, very rapid progress followed. Originally it looked much more nondescript than the four pairs of chromosomes in *Drosophila*. Torbjorn Caspersson noted that one could use certain fluorescent stains: he was looking for a way in which you could physically sort chromosomes, one from the other, to separate them. The staining provided rather peculiar banding patterns, not so different in principle from the bands in the salivary chromosomes of *Drosophila*, but not in as much detail, by an order of magnitude. But they do enable the cytogeneticist to look at a piece of chromosome and be able to identify it no matter where it appears, even if it has been the subject of a translocation, to spot it as having been a piece of X that's moved to chromosome number seven, or vice versa. And that has greatly accelerated progress in this field.

The first assignment of a gene to one of these cytological markers was in 1968 on the basis of a translocation; the Duffy marker was shown to be on chromosome number 1. At that time, nine linkage groups were known in the human, and we knew there were going to be 23 of them eventually, but it took one more finding to be able to do this with some expedition. With 23 chromosomes you have an exponential increase in the amount of labor that is needed to provide assignments of individual genes to individual chromosomes. The answer to that came from another quarter in yet another method of mapping, namely somatic cell fusion.

There was a lot of grumbling through the 1950s, including my own, about why we didn't treat somatic cells the way we have been treating microorganisms! We should be looking for sex where it had never been seen before. Eventually there were some exciting findings by Ephrussi and by Henry Harris (1970) in particular, that showed us the technology for doing that. One of the steps in the absorption and uptake of certain viruses in the mammalian cell involves the destruction of the cell membrane. By using irradiated suspensions of the Sendai virus, Harris was able to exploit this phenomenon, get enough destruction of cell surfaces that the cells could fuse with one another, and initiate the beginning of an artificially created quasi-sexual mechanism of bringing nuclei from different sources into the same cell to see whether they would interact.

They do in a number of interesting ways. In 1965, Harris and Watkins thought to do this between a mouse cell and a human cell; they were spectacularly successful. The whole set of mouse chromosomes is rather easy to tell apart from many of the human chromosomes, and because in these mouse-human cell hybrids the human cells tend to drop out, they just get lost. So it's not too difficult to establish the clones of cells derived from these fusions, which contain the full set of mouse chromosomes, plus one, and only one, human chromosome.

Then if you have the appropriate markers for detecting a gene, you can then tell whether a given marker is on that chromosome in the human set. That methodology very quickly amplified our knowledge of the assignment of human genes to chromosomes. By 1976, each chromosome had at least one marker gene. So a combination of banding, which enables us to identify specific pieces of human chromosomes, and this wonderful tool bench, on which one can follow the addition of one human chromosome at a time against the mouse karyotype background, has led to very rapid progress.

One can also locate markers on the chromosome by direct visualization. You can make a probe for the gene that you are seeking, starting with an RNA message extracted from active cells, and using reverse transcriptase to convert to c-DNA. Cloning the DNA of interest with the recombinant DNA methods into bacteria, you can get large quantities of DNA that is specific for a given function. Today, the PCR (polymerase chain reaction) (Mullis, 1990) makes it even easier to fabricate such probes. One of the most dramatic applications is *in situ* hybridization. If a fluorescent, or label, is attached to the probe, it can be used literally to light up the spot on the chromosome that displays homologous DNA (Lichter and Ward, 1990).

Using a combination of these techniques today, we now have extensive maps of the human genome. For many years, McKusick (1988) has been tabulating and listing about 5,000 markers; they are now beginning to be assigned to map positions. I guess there are five or six hundred that have been assigned, but maybe that was yesterday and there are a thousand today. Things are moving very very quickly.

The human chromosome set is comprised of 22 autosomes plus the X and the Y. If you examined just number 11, for instance, which is a medium-sized chromosome, you would find that it has about 120,000,000 base pairs or about 4% of the total genome. McKusick (1988) has recorded many interesting things that are now known to be located on that particular chromosome. His database emphasizes the genetic defects mapped thus far, using the variety of techniques that I have been describing. We have this level of extensive map data now for the whole chromosome set.

The ultimate map is the DNA sequence. The basic methodology consists of four experiments. To start, it is important to have some highly purified homogeneous material that is worth sequencing—for that, the recombinant DNA methodology is indispensable. Then you set up four different subsamples of that DNA for replication using one each of dideoxyadenylate or guanylate or cytidylate, or thymidylate (the A, G, C, T, of course, are the letters of the alphabet of which the DNA sequence is composed). By using the dideoxy derivative rather than the natural deoxy compounds in a small dose in a cocktail needed for synthesizing copies of the DNA to be sequenced, DNA synthesis is randomly interrupted when a dideoxy base is incorporated at that point in the enzymatic synthesis of copies of the DNA. The replication, then, goes no further. One can then sort the set of DNA molecules that have been synthesized according to the length of the piece: the long ones are on top, the short ones are on bottom. For instance, you can find out if a piece of X length appears as a band here in which case there is an A at that position because the synthesis of the DNA strand was terminated by a dideoxy nucleotide having been picked up instead of a normal deoxy nucleotide. As a control you see you don't have a band of this length in the G or the C or the T lane. And, similarly, for all the other bases. You find the sizes of the DNA fragments that end with that base. So by comparing the bands that you see in one and only one of the lanes, you can then literally write down the sequence of the DNA in that particular circumstance so that at this particular position, you had an A, you had a G, you had a G, you had an A, you had three Ts. And so on. Fundamentally, this is how sequencing is done.

While we can do this to stretches of hundreds to thousands of nucleotides, you also carry out the procedure on different DNA samples and find overlaps so that you can match and extend the results. You can put it all together and assemble the entire map of an organism. It has been done to the extent of about 200,000 bases, maybe more than that by this afternoon, with the method mentioned.

We still lack precise information on the human mutation rate per generation (Mohrenweiser *et al.*, 1989). The best estimates range from 10^{-5} to 10^{-6} per gene per generation for mutations of consequential effect: roughly 0.1 per genome, compared to an accumulated background of 10 to 100 times that level. Therefore, our evolutionary background would comprise 10 or 100 generations worth of new spontaneous mutations. Another way of putting those numbers is that it takes an average of 10 to 100 generations to weed out every deleterious new mutation that enters the gene pool. One guesses (very crudely) that every baby is born with about 100 genes exhibiting some new nucleotide substitution, almost all inconsequential. Regardless of this, we have to be very much concerned about aggravating the accumulation of genetic damage that is inherent in our evolutionary history, and which is unavoidable from natural radiation and cellular metabolism.

I turn now to the human genome project, which has emerged out of these scientific and technical advances (adapted from Lederberg, 1988).

DNA is a structure of formidable complexity. If unraveled from a single human cell, the three billion nucleotide pairs of DNA fulfill two meters of double helix. I should double that because it is only the haploid genome that has the three billion base pairs. If, as is widely assumed, about 1% of that total length is translated into protein structures, about 100,000 gene products will have to be accounted for. The ultimate reductionism would be to build an analytical factory that would complete the reading of all three billion nucleotides of human DNA as one technical exercise. A price tag of a few billion dollars is cited, perhaps less if there is prior investment in new technology to automate the task. Is it worth the cost? Undoubtedly.

Is it the wisest use of funds at that level of expenditure? I have very grave doubts. Part of my reservations have to with the style of research that it would encourage and part with the misunderstanding of what we need to learn in mapping the genome. By now we have profound information concerning a score or so of human proteins. Each of them is at least a life's work. At a modest ten million dollars per life's work, that would amount to a trillion dollars to gain detailed information about the full set of human genes. Obviously we cannot commit this large a sum, so we must make discriminating selections of targets before committing to the task.

About 100 human proteins are now discernable as agents of important biological activity. That number will soon grow to perhaps a thousand. That is roughly the map that McKusick prepared, although at present only a small fraction of human genetic disorders can be related to specific gene products, and not every one of them is of such compelling importance that you want to spend a hundred million dollars to identify it.

The biologically active genes and gene products should be the priority list for further inquiry. I am suggesting that we have sensible means by which to decide, out the 100,000 genes that we will eventually need to learn all about, that we can pick out a hundred of them, even a thousand of them, to analyze before the rest. Even this fraction of the total will take many lifetimes of work to try to plumb them to the bottom. Getting the DNA sequence connected with them is obviously absolutely essential, but to have merely sequenced the DNA is just the tip of the iceberg. We need to understand the functional aspects of the gene and the gene product.

To learn about a gene in depth we will have to look into detail of regulation, three-dimensional structure, genetic variability, within and between species, physiological interrelationships, and therapeutic applications. To pursue such inquiries will take much more than the engineering mentality that would apply the single methodology of DNA sequencing for a single sweep through the entire human genome. To make significant advances, we will need a sensitivity to the organism and the focus of expertise, even a fascination, with the particular gene system under scrutiny. My recipe is that we not overlook exploratory research, often best done in the context of natural historical observation. The field of view may be under the microscope or at the hospital bedside, as well as in the open countryside or the oceans.

With luck, our strategies will continue to be self correcting. As we are presented with samples of maps and sequences, individual chromosomes here and there, we will be better able to assess the value of coninuing the grand sweep, at a hundredfold greater costs, compared to the more focussed and diversified inspection of a thousand blossoms wherever they may appear.

Since this address was given on August 15, 1988, rapid leaps have been made in mapping many species, including *E. coli*, yeast, *Drosophila* and the human. Accordingly, the corresponding detail, now outdated, has been deleted from the current text. For extensive synopses and posters, see *Science* **250** (October 12, 1990)—chart at p. 262: a-p; *Science* **254** (October 11, 1991) at pp. 247-262; and *Science* **258** (October 2, 1992) at pp. 87-102.

LITERATURE CITED

Anderson, S., A.T. Bankier, B.G. Barrell, A.R. Coulson, M.H.L. Debruijn, J. Drouin, I.C. Eperon, D.P. Nierlich, B.A. Roe, F. Sanger, P.H. Schreier, A.J.H. Smith, R. Staden, and I.G. Young. 1981. Sequence and organization of the human mitochondrial genome. *Nature* **290**: 457-465.

Bernardi, G. 1989. The isochore organization of the human genome. *Ann. Rev. Genet.* **23**: 637-661.

Brewer, B.J. 1988. When polymerases collide—replication and the transcriptional organization of the *Escherichia-coli* chromosome. *Cell* **53**: 679-686.

Cairns, J., G.S. Stent, and J.D. Watson (eds.). 1966. *Phage and the Origins of Molecular Biology.* Cold Spring Harbor Laboratory of Quantitative Biology, Cold Spring Harbor, NY. 340 pp.

Harris, H. 1970. *Cell Fusion.* Clarendon Press, Oxford.

Lederberg, J. 1988. The second century of Louis Pasteur. A global agenda for biomedical research. Pp. 19-30 in *Molecular Biology and Infectious Diseases*, M. Schwartz, ed. Elsevier, Amsterdam.

Lichter, P., and D.C. Ward. 1990. Is non-isotopic *in situ* hybridization finally coming of age? *Nature* **345**: 93-95

McKusick, V.A. 1988. Mendelian inheritance in man. Catalogs of autosomal dominant, autosomal recessive, and x-linked phenotypes. Johns Hopkins University Press, Baltimore, MD.

Mohrenweiser, H.W., R.D. Larsen, and J.V. Neel. 1989. Development of molecular approaches to estimating germinal mutation-rates. 1. Detection of insertion deletion rearrangement variants in the human genome. *Mutation Res*. **212**: 241-252.

Mullis, K.B. 1990. The unusual origin of the polymerase chain-reaction. *Sci. Am.* **262**: 56.

Riley, M., and S. Krawiec. 1987. Genome organization. Pp. 967-981 in Neidhardt (1987).

Segall, A.M., and J.R. Roth. 1989. Recombination between homologies in direct and inverse orientation in the chromosome of *Salmonella*—intervals which are nonpermission for inversion formation. *Genetics* **122**: 737-747.

Singh, G., D.C. Wallace, and M.T. Lott. 1989. A mitochondrial-DNA mutation as a cause of Lebers Hereditary Optic Neuropathy. *New Eng. J. Med.* **320:** 1300-1305.

Sturtevant, A. H. 1965. *A History of Genetics*. Harper & Row, New York.

Yanofsky, C., and E.S. Lennox. 1959 Transduction and recombination study of linkage relationships among genes controlling tryptophan synthesis in *Escherichia coli*. *Virology* **8**:425-447.

Zinder N.D., and J. Lederberg. 1952. Genetic exchange in *Salmonella*. *J. Bact*. **64**: 679-699.

General Bibliographical Note:

Several monographic texts provide background detail and references for the molecular biology surveyed in this address. Particularly useful sources for the topics covered here are:

Darnell, James, Edwin Lodish, Harvey Franklin, and David Baltimore, 1990. *Molecular Cell Biology,* 2nd ed. Scientific American Books, New York.

Lederberg, J. (ed. in chief). 1992. *Encyclopedia of Microbiology.* Academic Press, San Diego, CA.

Neidhardt, F.C. (ed.). 1987. Escherichia coli *and* Salmonella typhimurium: *Cellular and Molecular Biology.* American Society for Microbiology, Washington, DC.

Neidhardt, F.C., M. Schaechter, and J.L. Ingraham. 1990. *Physiology of the Bacterial Cell. A Molecular Approach.* Sinauer Associates, Sunderland, MA.

O'Brien, S.J. (ed.). 1987. *Genetic Maps. A Compilation of Linkage and Reconstruction Maps of Genetically Studied Organisms,* 4. Cold Spring Harbor Laboratory, Cold Spring Harbor, NY.

Watson, J.D., N.H. Hopkins, J.W. Roberts, J.A. Steitz, and A.M. Weiner. 1987. *Molecular Biology of the Gene.* Benjamin/Cummings, Menlo Park, CA.

The history of molecular biology was comprehensively surveyed by: Judson, H. F., 1979. *The Eighth Day of Creation.* Simon & Schuster, New York. The early history of *E. coli* mapping is the subject of a memoir: Lederberg, J. 1987. Genetic recombination in bacteria: a discovery account. *Ann. Rev. Genet.* **21**: 23-46.

Special thanks are due to Professor John A. Moore, University of California at Riverside, for the photographic material on the Fly Room at Columbia University. See Moore, J. A. 1986. Science as a way of knowing. III. Genetics. *Am. Zool.* **26**: 583-747.

Frank Rattray Lillie
(1870 - 1947)

Edwin Grant Conklin
(1863 - 1952)

Determinants of Development: From Conklin and Lillie to the Present

Introduction by James D. Ebert

E. Newton Harvey began his biographical memoir of Conklin with these words: "**Edwin Grant Conklin**'s guiding principle in life was service, a rule of conduct which he expressed in an address at the dedication of the new brick building of the Marine Biological Laboratory at Woods Hole, Massachusetts in 1925: 'Our strongest instincts are for service; the joy of life is in progress; the desire of all men is for immortality through their work.'"

Conklin's "immortality," as perceived in the centennial year of the Marine Biological Laboratory, stems from three monumental contributions, published in the first 15 years of a career spanning more than 60 years. His monograph, "The embryology of *Crepidula*" was published in the *Journal of Morphology*, Volume 13, in 1897. Although it had been accepted by the editor C. O. Whitman in 1891, extensions and redrawing of figures prevented its appearance until 1897. The next comprehensive paper, which appeared in the *Journal of the Academy of Natural Sciences of Philadelphia*, Volume 12, in 1902, was titled "Karyokinesis and cytokinesis in the maturation, fertilization and cleavage in *Crepidula* and other Gasteropoda." The third large contribution, "Organization and cell lineage of the ascidian egg," was published in 1905, also in *Journal of the Academy of Natural Sciences of Philadelphia*, Volume 13. This work opened up a whole new vista in the study of "organ forming substances."

Conklin gradually turned his attention more and more to the application of biological discoveries to subjects of general interest, particularly with regard to humans, and particularly in the fields of heredity and evolution. His first book, *Heredity and Environment in the Development of Man*, was published in 1915. Harvey has written that Conklin always spoke of evolution as "The central theme of biology, the connecting strand on which all details of the science could be strung." Evolution was a theme in his publications beginning as early as 1896, and by the 1920s he had also begun to speak and write extensively on science and religion. He liked nothing better than to speak to a large public audience, the larger the better. At one time, according to Harvey, Conklin considered becoming a preacher but gave up the idea and was never ordained. However, he did receive what was known as a "local preacher's license." One of his students summed up his character as that of "a man of vigorous, definite, judicial, but amiable personality—a genuine sense of humor crops out unexpectedly to illuminate many a situation as in the famous remark—'Apparently the anti-evolutionist expects to see a monkey or an ass transformed into a man, though he must be familiar enough with the reverse process.'"

Conklin was born in Ohio to Nancy Hull and Dr. Abram Virgil Conklin. His father was a physician—a "horse and buggy doctor" of the 1870s. The family owned a farm and raised crops, cattle, and bees, and it is believed that this background of farm life led Conklin along a path of natural history. He attended a one-room school, worked on a farm, became the teacher in a similar school, and graduated from high school in Delaware, Ohio, in 1880. He then entered Ohio Wesleyan University where, under the influence of Edward T. Nelson, he collected shells of river mussels and snails. On graduation, Conklin took up a post in Rust University, a missionary college for Blacks in Mississippi where he taught, for three years (1885-1888) Latin, Greek, English, elocution, history, and all of the sciences.

In 1888, Conklin entered Johns Hopkins where he studied with W. K. Brooks, E. A. Andrews, H. Newell Martin, and others; Brooks was his principal (if at times reluctant) preceptor. It is clear that Conklin's career was shaped by his early experiences in Woods Hole, especially by his interactions with Professor E. B. Wilson. The two of them had been studying cell lineage, Wilson in an annelid and Conklin in a gastropod, and together they became convinced that in these two forms at least there was, as Conklin put it, "fundamental morphological significance in the cleavage."

Conklin's career took him back to Ohio Wesleyan for his first important teaching position. He then moved to Northwestern University, where he stayed for two years before becoming Professor of Zoology at the University of Pennsylvania. He remained at Penn for 12 years (1896-1908). Conklin flourished at the University and in the American Philosophical Society—the oldest of American learned societies—to which he was elected to membership in 1897. He was signally honored by election to the presidency of that society twice in his life, first in 1942-1945 and then in 1948-1952. As Harvey wrote, others had served the Society longer as president (Benjamin Franklin for 21 years), but no one but Conklin had been recalled for a second term, and that at the age of 85.

Conklin's list of honors and activities is truly extraordinary, but none were more significant to him than the MBL and the American Philosophical Society.

Although Charles Otis Whitman is rightly celebrated as the founding director of the Marine Biological Laboratory, the principal architect of the Laboratory, indeed of the scientific community of Woods Hole as we know it today, was **Frank Rattray Lillie**, an uncommonly gifted student of development at all levels—from cells to institutions.

A descendant of pioneer families of Scottish and English origin, Lillie was born in Toronto, Canada, on June 27, 1870. There was little more than a "trace" of science in his family background; his maternal grandfather, Thomas Rattray, was a nephew of the Scottish astronomer, Thomas Dick. Moreover, although by his own account Lillie leaned toward natural science in high school, it was not until his undergraduate days in the University of Toronto that he decided on a career in zoology rather than the ministry, which his family had expected. He was influenced in this decision by Professors R. Ramsay Wright and A. B. Macallum. It was the former who introduced Lillie to the MBL; in the summer of his graduation, 1891, Lillie attended the Laboratory's fourth session. It was there that he met Whitman. It was a meeting that was to shape his life, for he accompanied Whitman first to Clark University and then, in 1892, to the University of Chicago, where in 1894 he received the Ph.D. degree in zoology.

Except for an interval of six years (1894-1900), when he was at the University of Michigan and Vassar, Lillie's professorial life was intimately connected with the University of Chicago and the MBL. As Lillie himself put it, he served as an "understudy of Whitman being groomed as it were to be his logical successor." He assumed the chairmanship of Chicago's Department of Zoology in 1910, and served until 1931 when he resigned to take up the deanship of the Division of Biological Sciences, "amalgamating the preclinical departments with clinical departments and hospitals into a coherent medical school."

Lillie's capacity for leadership emerged even more rapidly in Woods Hole. He first came to the MBL in 1891, and from that year until 1946—55 years—he was present at the Laboratory every summer, first as an "investigator receiving instruction," and then as an instructor and later head of the embryology course. He served as assistant director of the Laboratory from 1900 to 1908 and as director from 1908 to 1925. Lillie truly shaped the Laboratory and made it a going concern.

Lillie's contributions in Woods Hole go far beyond the MBL. As early as 1925 he envisaged the establishment of other institutes in affiliation with the MBL, serving to "round out the scientific advantages of Woods Hole." With the backing of the National Academy of Sciences and the financial support of the General Education Board and the Rockefeller Foundation, Lillie was the prime mover in establishing the Woods Hole Oceanographic Institution (WHOI). It took him only six years from the initial steps to the full realization of his dream. He became the oceanographic institution's first president, serving from 1930 to 1939.

As I mentioned, the National Academy of Sciences played an important role in the establishment of WHOI. Lillie, in turn, played a key role in the Academy, serving on its National Research Council and its Division of Biology of Agriculture. He was president of the Academy from 1935 to 1939, and chairman of the National Research Council in 1935 and 1936.

Undoubtedly, Lillie, "the banker zoologist," used his own, and his wife's influential family (Crane) connections to good advantage. Even in those days there were concerns about interlocking directorates, as Ross Harrison observed while speaking about Lillie's account of the role of the National Research Council in abetting Lillie's campaign for funds for a new building. "Lillie's account of these events sounds somewhat naive in view of all of the invective that has been hurled at interlocking directorates, but it evidences a wholly honest spirit without fear of criticism for taking advantage of personal and official connections in such a good cause."

In research, Lillie led by example. He was actively engaged in original investigation for over 50 years (1891-1944). A listing of his contributions almost defines the experimental embryology of his times, except for the absence of any significant role in developmental genetics. Although he began his career with Whitman as a descriptive embryologist, in studies of cell lineage, Lillie thought of himself primarily as a student of the "physiology of development." Witness his contributions from the MBL on differentiation without cleavage and the physiology of fertilization, especially the "fertilizin" concept. His greatest contribution, however, came not from observations on marine embryos in Woods Hole, but from observations at the family farm near the village of Wheeling, Illinois, where, in his prized herd of pure-bred cattle, his attention was first drawn to the "freemartin," a barren female born co-twin to a normal bull calf. Lillie's mind had been prepared for this chance observation, for he had long been intrigued by the search for the primary cause of sexual differentiation. The freemartin, "a crucial experiment of nature," led Lillie to the discovery of embryonic sex hormones and to the exploration of their mechanism of action, a subject that predominated his research for the rest of his career, culminating in a final paper in 1947, the year of his death.

John Gurdon was educated at Eton and Christ Church, Oxford, and received has Ph.D. for work on nuclear transplantation in Amphibia. He held a postdoctoral position at California Institute of Technology (1961), was a visiting fellow at the Carnegie Institution (1965), and a lecturer at the University of Oxford (1962-1971). He has since worked at the MRC Molecular Biology Laboratory Cambridge (1972) and at Cambridge University (1985 to present).

His research in amphibian embryology has been recognized by several awards, including the Paul Ehrlich Prize (1977), the Charles Leopold Mayer Prize (1984), the Ross Harrison Award (1985), the Royal Medal (1985), and the Emperor Hirohito Prize (1987). Gurdon is a foreign associate of the National Academy of Sciences, the American Philosophical Society, and the Belgian Royal Academy.

Determinants of Development: From Conklin and Lillie to the Present

J. B. Gurdon

University of Cambridge

IT IS A VERY SPECIAL PRIVILEGE for me to be here this evening, having had the advantage of spending time a few years ago as a Lillie Fellow. I feel greatly honored to be invited back to participate in the MBL Centennial events, and especially to give this talk in honor of F. R. Lillie and E. G. Conklin, who are certainly two of the most famous and influential of American embryologists.

Let me start with the problem that both these distinguished scientists had as the center of their interest. This is the question of how the fertilized egg converts itself in a very short amount of time, often in a matter of hours but sometimes in a day or two, into a fully functional larva, in effect into a completely formed organism that mainly has to grow to complete the process of development.

I would like to illustrate this point with the same material that Conklin himself worked with. I refer here to the ascidian egg, which, within a day of fertilization, will convert itself into a fully formed larva. This larva is capable of swimming around, being well endowed with muscles right along the tail, and is the same size as the egg from which it derived. One important principle of development, illustrated by this example, is that there is no increase in material during this remarkable transformation.

The problem, therefore, is one of how the egg converts itself in this short amount of time into a completely organized larva with no influence from its environment. This is an amazing phenomenon and it occurs in nearly all organisms. It is this problem that fascinated Conklin and Lillie, and that I address here.

Background

We may start by asking how people, back in the 1700s, envisaged this process of early development. There were various theories at that time, and one that seemed to have considerable appeal was the idea that the egg or sperm had a miniature version of the complete organism inside it. Figure 1 shows what Dalenpatius thought he saw when he looked at human sperm in the 1700s (Needham, 1959). The figure shows two representations of sperm viewed with the facilities available at the time; note the hairstyle on the one figure and the neat disposition of the arms on the other.

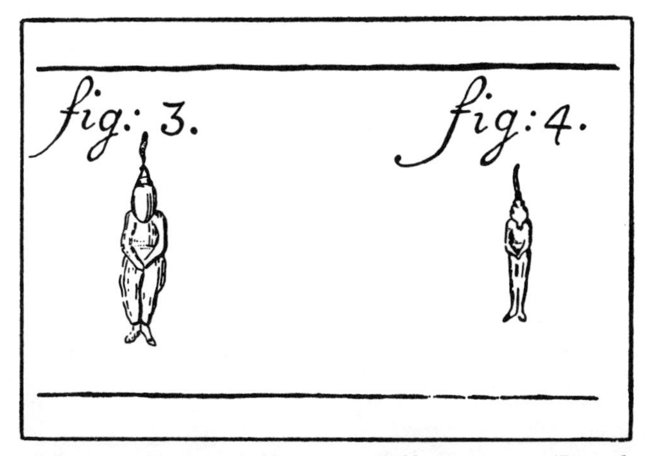

Figure 1. *Drawings by Dalenpatius of human sperm (Reproduced with permission from* A History of Embryology. *J. Needham. Cambridge University Press, 1959.)*

By the time of Conklin and Lillie, of course, people did not really believe that there was a miniature organism in the sperm or egg. Rather than looking inside an egg or sperm to see arms or a head, they were looking for regions of the egg that were in some way predisposed to form organs or tissues, such as muscle. Hence arose the concept of organ-forming substances that would, in due course, generate such structures as the muscle of ascidian larvae. This concept of organ-forming substances is one that we particularly now associate with E. G. Conklin. He made superb drawings of the ascidian *Styela* egg, which he chose because it had colored cytoplasm in certain regions. He found that the yellow-colored cytoplasm of the egg could be traced directly into the tail muscles of the larvae (Conklin, 1905). These original drawings, as a matter of fact, can still be seen in the Princeton Library.

I have been fortunate in being able to borrow from Dr. Richard Whittaker, who has himself carried out distinguished work for many years at the Marine Biological Laboratory, a superb series of photographs of this yellow plasm in a species closely related to that worked on by Conklin. The yellow-pigmented cytoplasm (Fig. 2) is present on the periphery of the unfertilized egg; at fertilization it moves to the lower part of the egg, and at each cleavage division, it becomes partitioned into cells that are direct ancestors of those that form the tail muscle. Conklin described in detail how a particular region of the fertilized ascidian egg is directly related, through the yellow pigment, to a particular type of cell differentiation or embryonic structure. This relationship was strongly emphasized in Conklin's writings, and is still one of the fundamental principles of development (Davidson, 1986). What we mean by this principle is that embryonic structures

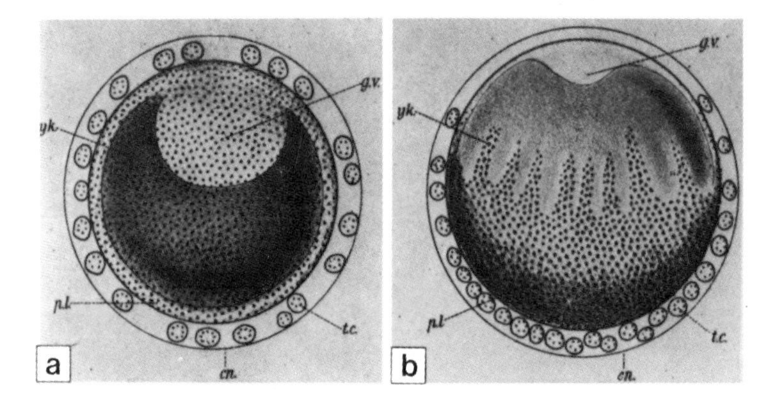

Figure 2. *Conklin's drawings of the egg of the ascidian,* Styela partita. *(A) Unfertilized egg, with a large nucleus (germinal vesicle, gv). The center of the egg contains gray yolk (yk), and a peripheral layer of yellow cytoplasm (pl). The egg is surrounded by a layer of test cells (tc) and a chorion (cn). (B) An egg 5 min after fertilization, showing the streaming of the peripheral yellow cytoplasm to the lower end of the egg (Reproduced from* J. Acad. Natl. Sci. Philadelphia, *Vol. 13, pages 1-119. E. G. Conklin. 1905.)*

and tissues are not themselves present in the undivided egg, either chemically or in a miniature form; nevertheless the fertilized egg contains a mosaic of substances that are directly related to future structures and cell-types of the embryo and larva.

Patterns of Cleavage

The next question I wish to discuss is whether the pattern of early cleavage divisions—the temporal and spatial order of early cell lineage—is important for the normal segregation or partitioning of cytoplasmic materials such as the yellow cytoplasm of ascidian eggs. At the time of Conklin and Lillie, many embryologists here at the MBL and elsewhere made detailed studies of cell lineage, trying to trace back—cell by cell—how a structure or tissue arose. This major preoccupation, I believe, was motivated by the thought that the pattern of the cleavage was quite important in development. Every species has a consistent pattern of cleavage that differs substantially from one kind of organism to another. It was, therefore, a reasonable guess that an important characteristic of development might be the pattern of early cleavages. We now believe that this is not the case. The order and arrangement of early cleavage divisions does not seem to be a causative factor in tissue differentiation. I would like to illustrate this point in two ways.

This gives me an opportunity to refer to an important experiment performed by F. R. Lillie. But first, I would like to reinforce a point that James Ebert makes in his introduction, namely that Frank Lillie was remarkable as both a scientist and an administrator; he was undoubtedly extremely successful in establishing the MBL. But that particular aptitude of combining science and administrative ability is one that is by no means universal, as illustrated by the case of a famous paleontological ichthyologist from this country, who knew by name almost every fossil fish that had been discovered. His success in this way was recognized in due course by an invitation to become president of a mid-western university, which he accepted. He then discovered, on moving, that it was diplomatic for him to learn the names of at least his most distinguished colleagues. Having made this effort, he used to complain that every time he learned the name of a colleague, he found that he forgot the name of a fish.

Returning now to Frank Lillie, the experiment that I wish to quote here was carried out on an annelid called *Chaetopterus*. Lillie discovered that if he treated an egg with a high concentration of salt, the egg would develop but the cells would not divide. The egg remained a single cell, and yet it differentiated to some extent like a normal embryo. Cilia formation and some degree of normal organelle arrangement occurred in a single cell (Lillie, 1902). Often quoted in textbooks, this remarkable result makes it quite clear that a pattern of cell division, and even cell division itself, is not necessary for at least the first stages of cell differentiation to take place.

The second piece of evidence that argues against a developmental role for the normal pattern of cleavage comes from the observation that certain eggs, which normally divide in a regular way, can be made to divide in a completely irregular way and yet give rise to an entirely normal individual. After nuclear transplantation in Amphibia, it is not uncommon for a "radial" eight-cell embryo to be formed, in contrast to the normal eight-cell embryo which has an upper tier of four cells resting on a lower tier of four vegetal cells. Nevertheless, the radially cleaved embryos with a single layer of eight cells are just as likely to form a normal larva and frog as are normal embryos.

We can conclude from the *Chaetopterus* and amphibian examples discussed, and from many other cases, that so long as a single cell egg is divided up into a large number of cells, the order and pattern of cleavage divisions by which this is achieved is unimportant. What matters is that organ-forming substances, such as the yellow plasm of ascidian eggs, should be segregated into cells corresponding in position in an early embryo to the position of the substances in the cytoplasm of an egg. Thus the views that Conklin and Lillie had, namely that organ-forming substances are critical for development, was in fact correct, just as we believe now. The only difference is that the pattern of cleavage by which they are divided up seems to be essentially irrelevant.

The Localization of Determinants or Organ-forming Substances in Eggs

Let me refer now to another very well-known organ-forming substance or material which is one of the best examples of egg cytoplasmic determinants. This is the pole plasm or germ plasm of insect eggs; this becomes localized in pole cells from which the gametes—or eggs and sperm of future adults—is derived (reviewed in Davidson, 1986). Ultraviolet irradiation of cytoplasm at one end of an uncleaved egg eliminates the formation of gametes, but has no other adverse effect; irradiated eggs form apparently normal adult flies, which, however, are sterile. There is a relationship between this particular bit of cytoplasm (the pole plasm) and fertility in this fly. This exemplifies the principle of an organ-forming substance being present in the egg, being partitioned out to pole cells, and helping to form particular kinds of cell types.

Having explained this general principle, let us see what more can be said about the ascidian muscle-forming substance. It is particularly appropriate to be able to do this here because much of the most distinguished work in this area has been done at the MBL. I shall quote two examples of work involving the yellow plasm of ascidian embryos. First, a very elegant experiment was done by William Jeffery (1982) concerning the localization of this yellow substance. An ionic calcium ionophore was attached to a glass rod and it was shown that wherever fertilized ascidian eggs touched this, the yellow plasm moves to that point. This is a striking demonstration of how the release of calcium will orientate the yellow plasm to one particular part of the egg, the whole process simulating a fertilization event, and the orientation of the yellow plasm in relation to the fertilization point.

The second piece of work concerning ascidian myoplasm comes from Richard Whittaker (1977, 1979). This is again a rather remarkable experiment in which he asked whether the localization of cytoplasm containing this yellow substance really commits the cells which receive it to form muscle. The answer is yes. He showed this by inhibiting cell division at a very early stage and by using a molecular marker, in this case acetylcholinesterase, as a measure of muscle-specific gene expression. The localization of this muscle-forming activity has taken place at the one-cell stage, and if the embryo is prevented from undergoing any more division, then there is already enough information in those cells to express muscle gene activity. This experiment completely excludes the idea that cell interactions are required for the expression of this organ-forming substance. We do not know yet what the substance is and that will, of course, take considerable time to find out.

Genetic Analysis of Cytoplasmic Determinants

Many laboratories are using genetics to try and identify organ-forming substances. But when was genetics first used? As far as I can determine, the first work was done in the 1920s involving a mutation in snails that causes left-handedness of the shell and body structure (Fig. 3). A remarkable genetic analysis by Boycott *et al.* (1931) in the

Figure 3. *Left-handed (sinistral) and right-handed (dextral) forms of a snail. The sinistral species (on the left) is* Neptunea contraria*; the dextral species (on the right) is* Neptunea antiqua. *The forms illustrated are the most common forms of each species, both of which have dextral and sinistral variants, respectively. (Photo by M. J. Ashby, University Museum of Zoology, Cambridge.)*

1920s showed that the handedness of a snail is determined by the genetic constitution of the mother. (As you might expect, Sturtevant and others contributed to this study.) The direction of coiling of the snail's shell depends on very early events of cleavage after fertilization and is determined by a maternally expressed gene. This work seems to me to mark the beginning of the genetic analysis of development.

Since that time, major advances have been made, most particularly in early *Drosophila* development. Several genes, and in some cases their products, have recently been identified which seem to determine the polarity and segmentation of the early embryo (see Akam, 1987, for a review). Commonly, it seems to turn out that the product of one gene, itself localized, acts by binding to, and activating, other genes. Cascades of gene expression of this kind lead to increasing refinements in the spatial organization of the early embryo.

What is not at all clear, so far, is how these early acting genes eventually elicit expression of other genes that confer tissue or organ specificity, such as muscle-specific genes of the kind referred to in ascidians.

Cell Interactions

I believe there are two fundamental principles about how an egg turns into an embryo. One depends on the localization of substances in the fertilized egg and their distribution to daughter cells in different parts of the embryo. The second fundamental principle involves an interaction between two kinds of cells that have inherited different substances from the egg. It is the combination of those two mechanisms that we now believe must account for the detailed formation of an embryo.

Turning now to the cell interaction side of embryology, it is interesting, again, to look back historically to see how this field arose. I am referring now to the process called embryonic induction. This describes any case where one (responding) cell is made to change its direction of differentiation by the influence of another (neighboring) cell. This is different from processes of the kind that Lillie discovered in his freemartin experiments (Lillie, 1916). The conclusion from those experiments was that there are circulating substances—hormones—which influence some aspects of differentiation. A hormone spreads throughout the whole body and influences any cell in the body that can respond to it. Embryonic induction happens at a much more localized level, such that only cells very close to the source of these substances actually respond. The phenomenon of embryonic induction was actually discovered by Spemann (1901), and also by Lewis (1904) in this country. It is of some interest to look at the famous textbook by E.B. Wilson last published in 1925 (it was recently reprinted). There is, in fact, no mention of embryonic induction in that text. It is remarkable that these cell interactions seem to have been hardly known or at least not taken seriously by people working here in Woods Hole, which was well established as a center of developmental biology in the time of Wilson and Lillie.

Are cell interactions really important? Is this a fundamental cell mechanism of development? There is increasing evidence that this is so, at least for the vertebrate animals. I shall now limit my comments to Amphibia, because this is the material on which embryonic induction work has so far advanced furthest.

Figure 4. *A larva of the amphibian,* Xenopus laevis, *three days after fertilization.*

Muscle-forming Induction in Amphibia

Figure 4 is a picture of an amphibian larva, which can be compared to an ascidian larva. The former is much larger, but the two have several features in common. The major component of the amphibian larva, as with the ascidian, is body muscle, but it also has eyes, an attachment organ, a heart, and a nervous system. All of these components arise by embryonic induction. Not one of these could be formed were it not for the fact that cells of different kinds interact with each other.

In the case of Amphibia, it is clear that the major organ systems all depend on embryonic induction events. We now know that many of these inductions, such as lens differentiation, are much more complicated than seemed to be the case to Lewis and Spemann over 80 years ago. Lens induction requires at least three different cell interactions, not one of which is sufficient to activate lens-specific genes (Jacobson, 1966; Grainger *et al.*, 1988). For analysis, it is essential to select the simplest example of embryonic induction; this is probably the interaction between animal and vegetal pole tissues of a blastula, as a result of which body muscle is formed. This earliest induction in amphibian development initiates a chain of subsequent inductions, as summarized in Figure 5. Cells of the animal hemisphere of a blastula are competent for a limited period of a few hours to be induced to form mesoderm; the mesoderm subsequently forms

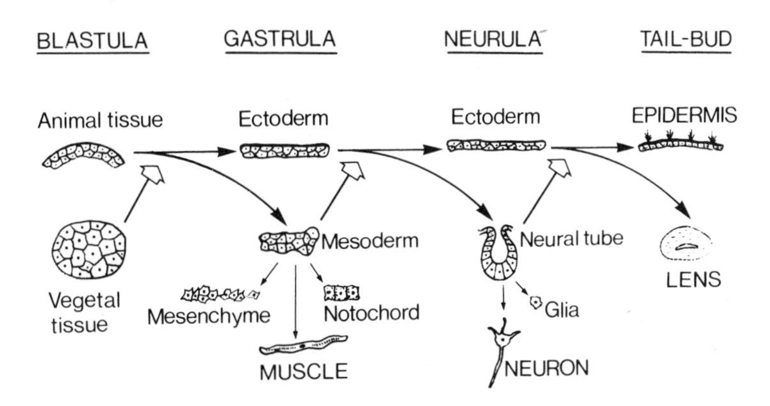

Figure 5. *A diagram of the sequential inductions by which most new cell-types originate in amphibian development. Solid arrows indicate directions of cell differentiation. Open arrows indicate inductive influences. Cell-types for which good molecular markers, expressed in embryos, are available are indicated in capital letters. (Reproduced with permission from* Trends in Genetics, *Vol. 5, pages 51-56. J. B. Gurdon* et al. *1989.)*

muscle, backbone, and part of many other internal organs. If the animal pole cells are not induced by vegetal cells, they soon lose that capacity, and acquire a competence to be induced by the newly formed mesoderm to form the nervous system. Some animal cells do not receive any inductive influence, because they are too far from the source of inducer; these cells then follow the uninduced direction of differentiation into epidermis.

To analyze any biological process, it is essential to have a simple assay. A group of animal cells at the top of the embryo, which are naturally pigmented, are placed on top of the white vegetal cells, which are inducers. Within 2 hours, the responding animal cells, which are nearest the vegetal cells, have been diverted from the path of epidermal differentiation towards muscle differentiation (Nieuwkoop, 1969). There is a sensitive and quantitative biochemical procedure by which the earliest transcripts of a muscle-specific gene can be recognized only a few hours after the start of induction (review by Gurdon, 1987). There are two ways forward from a situation like this. One is to analyze the process starting with an inducer substance and work forward until one understands how muscle is formed, and the other is to take a specific response such as muscle gene activation and try to work backwards to find out how it came about. It turns out that both of these directions of investigation are now proving useful.

Considering the first approach, it has recently been discovered that mammalian growth factors are effective as inducers (Slack *et al.*, 1987) and the most hopeful work in my view is that of Doug Melton, who has identified a particular gene called Veg1 whose protein product seems to have almost all the qualities expected of inducer substances (Weeks and Melton, 1987). This gene is especially interesting because its products are localized to the vegetal part of the embryo just as the ascidian yellow plasm is localized to an equivalent position in the ascidian embryo. Melton has shown that Veg1 gene transcripts are uniformly distributed throughout the cytoplasm of a growing oocyte, and only later become localized to the vegetal pole. This justifies the important conclusion that a maternal gene can make a product that is initially widely distributed, but in the course of the formation of the oocyte moves to one end of that cell to become localized. Furthermore a particularly important result is that direct transcripts of this gene injected into oocytes become appropriately localized (Yisraeli and Melton, 1988); this is a way of finding out the sequence of a gene's product which causes localization, and can hence lead on to an analysis of the mechanism of localization.

In conclusion, to work from the inducer forward looks very encouraging, but the problem still remains of how to relate that inducer to the eventual activation of a muscle gene.

Analysis of Embryonic Induction from a Specific Gene Activation Backwards

In the experimental system outlined above, there is a period of only 9 hours from the first contact of animal with vegetal cells and the first detection of muscle gene transcripts. The following comments summarize attempts to piece together the events that take place during these 9 hours (Gurdon *et al.*, 1989). A key feature of this analysis is to purify a muscle actin gene, which is turned on very early in Amphibia as it is in ascidian development, and then to determine which part of the gene is needed for it to respond to the inducer.

It has been established that 400 nucleotides of a muscle actin gene 5' to the start of transcription is sufficient for a strong response to induction, even if the whole of the coding region and introns of the cardiac actin gene are replaced by another gene. Within these 400 bases, a particular 10 base sequence around -85 with respect to the transcription start seems especially important, and proteins that bind specifically to this sequence can be extracted from cells. Current work is aimed at trying to identify the nature of these proteins. This, in turn, opens up the possibility of determining whether these and other proteins needed for the response to induction arise by modification of preformed proteins, by translation of maternal messages, or by the new transcription of a gene. All of this takes place some seven to eight hours after the inducer has been applied. So what happens before that? Can we fill in some of the preceding events? It turns out that there are two very different kinds of cell interactions involved at that time.

One is seen early on, when inducer substances move freely from vegetal to animal cells and initiate induction. This whole process can occur in the absence of physical contact between cells.

The next kind of cell interaction is different, and is termed "the community effect," because it seems that responding cells must interact with each other. This interaction proceeds only if reacting cells are in close contact with each other. The existence of this effect was demonstrated by placing responsive animal cells, either as a monolayer or as a three-dimensional group, into a sandwich of vegetal cells. The animal cells arranged as a monolayer did not activate muscle genes, whereas those reaggregated as a solid tissue did. The principle here seems to be that similar responding cells have to interact with each other as a community. The community effect, as it is called, is seen as one in which like cells have to be doing the same thing at the same time to complete a response. One might wonder how this could happen. If each cell were to release a substance that only reaches the necessary concentration when there are many cells together thereby surpassing some threshold, then this phenomenon could be understood.

The function of the community effect is, in my view, to give a homogeneous collection of cells doing the same thing at the same time. And this happens very commonly with muscle, nerve, and all those kinds of differentiation discussed above. What it would do, in effect, is to eliminate the floating voter or the cell that does not know in which way to differentiate.

Summary and Conclusions

I spoke initially about the Conklin/Lillie concept of determinants and moved on to cell interactions, thus referring to the two major mechanisms used by all animals to create differences between cells. In the time of Conklin and Lillie, nearly all of the emphasis in experimental embryology was on the localization, and distribution in early cleavage, of organ-forming substances or determinants. Since then the importance of cell interactions and embryonic induction has become much more obvious, especially in the vertebrates. However, I have tried to point out that both mechanisms are used in the invertebrates as well as in the vertebrates.

I have emphasized experimental work with Amphibia, partly because of my own familiarity, and partly because it illustrates especially well the cooperative usage of both major developmental mechanisms. In Amphibia, localized substances already exist in the fertilized egg, creating animal and vegetal pole cells. Thus an interaction between these two cell types progressively amplifies the range of cell-type differences. It is pleasing to note that the fundamental concepts of Conklin and Lillie are every bit as valid today as in their time, some 80 years ago.

Let me finish with a comment on the importance of the Marine Biological Laboratory as an international center of embryological research. This laboratory continues to be a major international focus of embryological research, and I would like to add my felicitations to the MBL's first century while looking forward to its second century.

LITERATURE CITED

Akam, M. E. 1987. The molecular basis for metameric pattern in the *Drosophila* embryo. *Development* **101:** 1-12.

Boycott, A. E., C. Diver, S. L. Gasstang, and F. M. Turner. 1981. Inheritance of sinistrality in *Limnaea peregra. Phil. Trans. Roy. Soc. B.* **219:** 51-131.

Conklin E. G. 1905. The organization and cell lineage of the ascidian egg. *J. Acad. Natl. Sci. Philadelphia* **13:** 1-119.

Davidson, E. H. 1986. *Gene Activity in Early Development,* 3rd ed. Academic Press, New York.

Grainger, R. M., J. J. Henry, and R. A. Henderson. 1988. Reinvestigation of the role of the optic vesicle in embryonic lens induction. *Development* **102:** 517-526.

Gurdon, J. B. 1987. Embryonic induction—molecular prospects. *Development* **99:** 285-306.

Gurdon, J. B., T. J. Mohun, M. V. Taylor, and C. Sharpe. 1989. Embryonic induction and muscle gene activation. *Trends in Genetics* **5(2):** 51-56.

Gurdon, J. B. 1988. A community effect in animal development. *Nature* **336:** 772-774.

Jacobson, A. G. 1966. Inductive processes in embryonic development. *Science* **152**: 25-34.

Jeffery, W. R. 1982. Calcium ionophore polarizes ooplasmic segregation in ascidian eggs. *Science* **216:** 454-547.

Lewis, W. H. 1904. Experimental studies on the development of the eye in Amphibia. *Am. J. Anat.* **3:** 505-536.

Lillie, F. R. 1902. Differentiation without cleavage in the egg of the annelid *Chaetopterus*. *Arch. Entwicklungsmech. Org.* **14:** 477-499.

Lillie, F. R. 1916. The theory of the free-martin. *Science* **43:** 611-613.

Needham, J. 1959. *A History of Embryology.* Cambridge University Press. 205pp.

Nieuwkoop, P. D. 1969. The formation of the mesoderm in urodelean amphibians, induction by the endoderm. *Wilhelm Roux Arch.* **162:** 341-373.

Slack, J. M. W., B. G. Darlington, J. K. Heath, and S. F Godsave. 1987. Mesoderm induction in early *Xenopus* embryos by heparin-binding growth factors. *Nature* **326:** 197-200.

Spemann, H. 1901. Uber Korrelationen in der Entwicklung des Auges. *Verh. Anat. Gesel. 15 Vers Bonn.* (*Anat. Anz.* **15**): 61-79.

Weeks, D. L., and D. A. Melton. 1987. A maternal mRNA localized tothe vegetal hemisphere in *Xenopus* eggs codes for a growth factor related to TGF-b. *Cell* **51:** 861-867.

Whittaker, J. R. 1977. Segregation during cleavage of a factor determining endodermal alkaline phosphatase development in ascidian embryos. *J. Exp. Zool.* **202:** 139-154.

Whittaker, J. R. 1979. Cytoplasmic determinants of tissue differentiation in the ascidian egg. Pp. 29-51 in *Determinants of Spatial Organization*, S. Subtelny and I.R. Konigsberg, eds. Academic Press, New York.

Wilson, E. B. 1925. *The Cell in Development and Heredity*, 3rd ed. Macmillan, New York.

Yisraeli, J. K., and D. A. Melton. 1988. The maternal mRNA Vgl is correctly localized following injection into *Xenopus* oocytes. *Nature* **336:** 592-595.

Albert Szent-Györgyi
(1893 - 1986)

Albert Szent-Györgyi: "To see what everyone has seen and think what no one has thought"

Benjamin Kaminer

Boston University School of Medicine

Benjamin Kaminer received his medical degree in 1946 from the University of Witwatersrand in Johannesburg, South Africa, and after clinical internships went to London in 1949 for further training and research experience. On returning, he joined the physiology department at Witwatersrand. In 1959, a Rockefeller Fellowship enabled him to join Albert Szent-Györgyi's Institute for Muscle Research at the Marine Biological Laboratory, where he worked year-round until 1969. Since then he has returned to the MBL every summer. He has served as an MBL trustee and on the Executive Committee. He is currently chairman of the physiology department at Boston University School of Medicine.

SHORTLY BEFORE HIS 93RD BIRTHDAY, Albert Szent-Györgyi, threatened with a loss of financial support from a certain cancer foundation, decided with a renewed spark of vigor and determination to maintain his laboratory at the Marine Biological Laboratory. He publicly pronounced his intention, but died shortly thereafter on October 22, 1986. This last display of determination, defiance, and daring, but with a maintained sense of humor, epitomizes the man we honor as one of the "Greats" of the MBL at our centenary. On being presented with the Nobel Prize in 1937, Albert Szent-Györgyi was referred to as the "new conquistador from Szeged," who "conquered new ground by intuitive daring and skill" with a "clear vision for essentials." (Hammerstein, 1937) Much of his early life is portrayed in an autobiographical sketch (Szent-Györgyi, 1963).

Albert Szent-Györgyi von Nagyrapolt, born in Budapest on September 16, 1893, decided as a young boy to become a scientist. His father, a semifeudal landowner preoccupied with farming, cared little about his intellectual development, in contrast to his mother Josefine who inculcated a love of science, art, and music in the family. Her father and grandfather had been professors of anatomy and physiology in Vienna, and her brother Mihaly Lenhossek was professor of anatomy in Budapest. Lenhossek, a noted histologist at the time, and an influential member of the family, always regarded Albert as a dull child and did not encourage him to pursue a scientific career. But Albert later became a voracious reader, and when he graduated from high school with top marks his uncle agreed to his admission to medical school. As a student he immediately embarked on research, published papers on the histology of the anus and the eye (*sic*) but soon became disenchanted with morphology. In his third year at medical school he turned to research in the physiology department. His studies were interrupted by the outbreak of World War I when he was conscripted into the army. After three years, having received the Silver Medal for valor, Szent-Györgyi, disgusted with the brutality of war, was utterly disillusioned and desperately wanted to get out of the army to continue with research. He saw no way out, so he shot himself in the arm. He returned to medical school, obtained his M.D. degree, and was recruited into a military bacteriology laboratory. There he soon landed in conflict with his superiors by objecting to risky experiments on Italian prisoners of war. He was sent to serve in swampland rife with malaria in northern Italy. Fortunately, the war ended a few weeks later and he became an assistant in a pharmacology department in a newly founded university in Pozsony that, within months of his appointment, was handed over by the Versailles Treaty to Czechoslovakia.

From Prague to Holland

Forced to leave, Szent-Györgyi went to Prague to learn electrophysiology from Armin von Tschermak. He proceeded to Berlin to work with Michaelis and then to Hamburg at the Institute for Tropical Hygiene where he met W. Storm van Leeuwen. Van Leeuwen invited him to take a position in pharmacology in Leiden, where he spent two years before moving to Hamburger's laboratory in Groningen. In Michaelis' laboratory, Szent-Györgyi was introduced to physical chemistry and studied electrolytes and their effects on proteins (*i.e.*, the agglutination process). He also published a paper on an experimental study of the molecular theory and determination of the value of Avagadro's number. In Leiden, his investigations of drug action included the effect of colloids on drugs.

In Groningen, Szent-Györgyi (as second author with Brinkman) published a number of papers on hemolysis and on the permeability of collodium membranes. In 1924, at the age of 31, he embarked on one of his most significant studies. He became interested in the process of tissue oxidation. At that time, Otto Warburg had discovered cytochromes that act as catalysts in tissue oxidation. Wieland, also working on cell respiration, discovered dehydrogenases that removed hydrogen from molecules to eventually link with oxygen, forming water. A debate arose between Warburg and Wieland as to whether cell respiration was due to activation of oxygen or hydrogen. Szent-Györgyi, having a hunch that both were important, did a very simple experiment, which led the way to the resolution of the conflict; he added cyanide to minced pigeon breast muscle and stopped respiration. He then added methylene blue, a hydrogen acceptor, which restored it. In Groningen, he continued his research on the metabolism of minced pigeon breast muscle, a preparation that he introduced to biochemists. Later, in Hungary (from 1930 to 1936), he elucidated the catalytic role of C4-dicarboxylic acids in the oxidation process which laid the foundation, despite conceptual errors, of the well-known Krebs cycle.

He was also interested in plant respiration "being convinced that there was no basic difference between man and the grass he mows." When one adds peroxide to peroxidase and benzidine the solution turns blue owing to the oxidation of benzidine. Szent-Györgyi performed such an experiment with plant juice instead of pure peroxidase, and this caused a very slight delay in the appearance of the blue color. He argued that the juice reduced the oxidized benzidine causing a time delay in the color reaction. He proceeded to extract the reducing agent from various vegetable sources. He imagined that this reducing agent would also be present in the adrenal cortex because in Addison's disease, when the cortex is destroyed, pigmentation occurs in the skin. He equated this phenomenon with the blackening of a cut potato, which is associated with the oxidation of phenols after the loss of a reducing agent. His analogy was obviously wrong, but he nevertheless found the reducing agent in small amounts in the adrenal gland. He used this story to illustrate his way of working: "I make the wildest theories, connecting up the test tube reaction with broadest philosophical ideas, but spend most of my time in the laboratory playing with living matter, keeping my eyes open, observing and pursuing the smallest detail. The theories serve to satisfy the mind, prepare it for an accident and keep one going. I must admit that most of the new observations I made were based on wrong theories. My theories collapsed, but something was left afterwards."

The Road to Cambridge: Ignose or Godnose

His publication on the oxidation—the blackening—of a cut potato had special significance for his future. At that time, Szent-Györgyi was in great despair. His boss, Hamburger, had died in 1926, and Szent-Györgyi did not know where to go or whether he would ever be able to continue with his research. His wife and small daughter returned to Hungary, and he decided to attend the International Physiological Congress in Stockholm. Szent-Györgyi's work on blackening of potatoes had caught the attention of Sir Frederick Gowland Hopkins, who referred to it in his presidential address at this congress. Szent-Györgyi introduced himself to Hopkins, who immediately invited him to Cambridge where he was head of the biochemistry department. With the support of a Rockefeller Fellowship, Szent-Györgyi soon arrived in Cambridge, which he thereafter regarded as his scientific home. There he continued the work he had begun in Groningen and crystallized the reducing substance from fruit, vegetables, and the adrenal gland. The molecule appeared to be a 6-carbon sugar, but he was ignorant of its chemical structure. So with tongue in cheek he combined the term ignorance with" -ose" (for sugar) and named it "Ignose" in the article he sent to the *Biochemical Journal*. The editor rejected the title so Szent-Györgyi renamed it "Godnose" which was also rejected. Harden, the editor, proposed the term hexuronic acid, a name that appeared in the published paper. This name was shown later to be inappropriate; the substance turned out to be Vitamin C. Because Szent-Györgyi's best yield at the time was from adrenal glands, he arranged to work in Kendall's laboratory at the Mayo Clinic, where he could obtain adrenal glands in abundance from the slaughter house in St. Paul, Minnesota. After a year's leave of absence, he returned to Cambridge with 25 grams of "hexuronic acid," most of which he sent to Haworth, a renowned carbohydrate biochemist in Birmingham, England, for further analysis. In Cambridge he found time to embark on other research projects and with

Drury published in 1929 on the identification and crystallization of adenylic acid from tissue extracts . The authors showed that adenylic acid and adenosine influenced the conduction of heart muscle and decreased blood pressure of animals by dilating vascular smooth muscle. These findings have now been rediscovered, and adenine compounds are being recognized as transmitters of cellular signals, and as important regulators of coronary blood vessels. They are also being used in the treatment of cardiac arrhythmias. For his research on "hexuronic acid" Szent-Györgyi received the Ph.D. degree from Cambridge University in 1930.

Return to Hungary: Nobel Prize

At the invitation of the Hungarian Minister of Education, Szent-Györgyi returned to his homeland to help modernize Hungarian science and accepted the chair of Medical Biochemistry at the University of Szeged. He continued with his research on oxidation and then became interested in a yellow substance he termed "cytoflave," which fluoresced. Without a spectroscope his investigations were limited. The substance was similar to Warburg's yellow enzyme later named riboflavin. On investigating lactate dehydrogenase, Szent-Györgyi isolated a coenzyme, a nucleotide that Warburg, working in a similar direction, identified as a pyridine nucleotide. Luck knocked at Szent-Györgyi's door in 1931, when Joseph Svirbely, an American of Hungarian descent, arrived in Szeged. Svirbely had done his Ph.D. thesis research with Charles King at the University of Pittsburgh. There he had isolated Vitamin C from lemons and had experience in assaying substances that prevented or alleviated scurvy in guinea pigs. Szent-Györgyi gave him the small amount of "hexuronic acid" he had left for assaying, and its effect was similar to that of Vitamin C isolated from lemons. After reproducing the exciting results, they agreed that Svirbily should inform King, which he did by letter in March 1932. King and Waugh (1932) published a note in *Science* magazine on April 1 on the identity of the chemical properties of Vitamin C from lemons and hexuronic acid. Sixteen days later, *Nature* magazine reported the discovery by Svirbily and Szent-Györgyi (1932). Szent-Györgyi had, of course, crystallized and characterized hexuronic acid whereas King and Waugh had simply repeated some tests showing similarities between their crystals obtained from lemon extracts and hexuronic acid. In any event, Szent-Györgyi had communicated their results to the Hungarian Academy of Science twelve days before the appearance of the note by King and Waugh.

Many conflicts, particularly in the press, and animosity arose over the claims of priority for the discovery of Vitamin C and the awarding of the Nobel Prize to Szent-Györgyi in 1937. The prize, it should be noted, was awarded only after Szent-Györgyi (1935 and 1936) had published his important papers on the role of fumaric acid in oxidation. Contrary to the popular belief that the prize was awarded for the discovery of Vitamin C, the Nobel Committee's citation on the award to Szent-Györgyi reads "for his discoveries in connection with the biological combustion processes, with special reference to Vitamin C and the catalysis of fumaric acid" (Hammerstein, 1937). Szent-Györgyi's Nobel Prize acceptance lecture in Stockholm was entitled "Oxidation, Energy Transfer and Vitamins." In a ten-page publication of this lecture, Szent-Györgyi devotes two pages to Vitamin C, dealing only generally with the story of how he got onto the idea of the presence of a reducing substance in vegetables and the adrenal gland and the events thereafter; he did not discuss its chemistry. Friends and supporters of King therefore have no basis for their view that he should have shared the Nobel Prize with Szent-Györgyi. The discovery of Vitamin C was not the primary reason for Szent-Györgyi receiving the Nobel Prize. However the press caught onto the words "Vitamin C" in the announcement of the award, popularized it, and thus contributed significantly to Szent-Györgyi's fame. No one, not even Szent-Györgyi, bothered to correct this erroneous impression. Max Perutz (1988), who investigated thoroughly the files of the Nobel Committee, points out that in 1934 the Committee declined to make an award for the discovery of Vitamin C. The Committee recognized the importance of Szent-Györgyi's identification of hexuronic acid as Vitamin C, but too many other investigators, in their opinion, had contributed to the vitamin part of the story to enable them to give a joint award. (Joint awards are limited to three persons.)

Meanwhile, Szent-Györgyi was concerned about his limited supply of hexuronic acid. Suddenly he realized that although he had used a great variety of vegetables, each yielding small amounts of crystals, he had not yet tried to extract Vitamin C from Hungarian fresh peppers (paprika). This realization came to him one night while eating paprika for dinner. He immediately ran to the laboratory and soon obtained a high yield. By the end of the paprika season he had crystallized half a kilogram of the material. He sent the material to investigators all over the world, including Haworth, who determined the precise structure of the molecule; they renamed it ascorbic acid. Szent-Györgyi continued to produce large amounts of Vitamin C and to distribute it widely to anyone wishing to study the molecule and for clinical experimentation in areas of the world where starvation was rife.

Actin and Myosin

Szent-Györgyi's fascination with energy sources driving the living process led him to his studies on muscle, a tissue that undergoes remarkable energy transformations between the states of relaxation and contraction. Between 1940 and 1944, during the turbulence and hardships of the Second World War, Szent-Györgyi and his associates, notably Ilona Banga and F. Bruno Straub, laid the foundation for our current understanding of the molecular mechanisms of muscle contraction. In 1939, Engelhardt and Ljubimova, two Soviet scientists, had made the important observation that myosin (extracted from muscle according to the procedure of Kuhne published about 100 years previously) showed an ATPase activity. With Banga, Szent-Györgyi showed that this myosin, extracted with high KCl over a 24-hour period (myosin B), was more viscous than preparations extracted over a 20-minute period (myosin A). They argued that the viscous preparation was a complex of two proteins. Threads made from the viscous preparation in a low ionic strength solution could be made to shorten by adding a water extract of muscle. This observation, according to Szent-Györgyi, was one of the most exciting moments in his life. Experiments soon revealed that the essential ingredient causing shortening was ATP in the presence of Mg and K. The major component in the muscle extract was purified and "crystallized" by Szent-Györgyi, and he continued to call it myosin. The second protein was purified by Straub and was named actin. The pure myosin had a low ATPase that was activated by actin. Threads made from pure myosin could not be induced to shorten with ATP, whereas threads made from a combination of the two proteins called actomyosin did shorten. Actomyosin also superprecipitated in association with ATP hydrolysis. Szent-Györgyi, with remarkable insight, regarded the interaction of actomyosin with ATP as the fundamental process in muscle contraction. The discoveries were briefly written up, and a few hundred copies of three volumes of "Studies from the Institute of Medical Chemistry, University of Szeged" were printed in the local printing shop. To ensure against their loss during the war, Szent-Györgyi managed to send the manuscripts to Stockholm, and they were published in *Acta Physiologica Scandinavica* (Szent-Györgyi,1945).

Szent-Györgyi thus pioneered the approach of reconstituting partially a functioning structural system from isolated components. His basic concept of muscle contraction, which became more widely known after the publication of his booklet on the *Chemistry of Muscle Contraction* in 1947, was, however, seriously criticized by Kenneth Bailey (1947), a respected muscle biochemist, in a review of the book in *Nature*. Bailey considered it a "rather novel and daring approach to the problem of physiological contraction," but among various criticisms he wrote, "In all that concerns the properties of proteins there are serious misconceptions." "...To transpose this system [actin-myosin-ATP] to muscle itself is not to offer an explanation how muscle contracts." "...It [the book] should not be recommended to uncritical and over receptive students." Fortunately, many were more than receptive to the exciting ideas in this booklet (and in the others that soon followed), which opened the frontiers for our current understanding of muscle contraction.

Soon after commencing his studies in muscle, Szent-Györgyi published an article in 1940 "On protoplasmic structure and function" in which he refers to his profound conviction that basic processes of life will be the same in all cells and organs. After discussing muscle and on considering protein structure in general he wrote, "we will come to the conclusion that wherever the organism wants to build a solid structure it will resort to a rod-shaped protein and wherever it needs a certain mobility it will apply the globular form" (Szent-Györgyi, 1945). At that time most proteins were found to be globular except those in special tissues such as muscle and fibrous tissue and in hair and fibrin. Szent-Györgyi, contrary to the ideas at the time, felt that all cells contained rod-like filamentous proteins but that these were lost by the biochemist because they were not readily soluble. He wrote, " a protein fraction analogous to myosin [*i.e.*, myosin B containing actin] should be found in any cell." He proceeded to describe the treatment of numbers of tissues with solutions containing high salt concentrations at varying alkalinities and also with solutions of urea aimed to break hydrogen bonds. The extracts had a sticky consistency and, in a polarizing microscope, showed double refraction of flow suggesting the presence of rod-shaped molecules similar to those in crude extracts of myosin B. He extracted these proteins from kidney, liver, brain, nerve, mammary gland, parotid, lymph gland, whole embryos, *Rous sarcoma* and *Ehrlich carcinoma*. Hence, Szent-Györgyi, in advance of his time, anticipated the discovery of myosin and actin in non-muscle cells. This concept, and Szent-Györgyi's findings, have somehow been missed in the frequent reviews on the history of the discovery of these proteins as part of the so-called cytoskeleton and their significance in cellular organization and function.

At this early stage of great productivity on muscle proteins in his laboratory, Szent-Györgyi's restless mind was apparently not satisfied with a molecular explanation of muscle contraction; he was looking further ahead. His intuitive ("Dyonisian") approach led to his speculations on energy levels in proteins as expressed in his Koranyi lecture in 1941 and he predicted "a new period in biochemistry, taking this science into the realm of quantum mechanics." "Prof" as Szent-Györgyi was affectionately known, always stimulated his students and associates with exciting new ideas and fascinated them with his formulation of simple questions and concepts on the most complicated processes. During this period, he organized joint seminars with theoretical physicists and generated discussion and speculations on proteins being semiconductors and on the role of electrons in biological processes in general. He was already paving the way for his eventual digression from muscle research in the mid-fifties at the Institute for Muscle Research, which he established in 1947 at the Marine Biological Laboratory in Woods Hole.

At the MBL, Prof finally overcame the criticisms of his basic theory of muscle contraction. His extracted preparations of actomyosin, as already mentioned, superprecipitated (which he claimed was analogous to contraction) in association with the hydrolysis of ATP, and in the form of threads actomyosin could shorten (as does muscle), but could not produce tension. So his critics argued that Prof's extrapolations to living muscle were far fetched. He then conceived the brilliant idea (how it came to him he claimed not to know) of extracting bundles of rabbit psoas muscle in 50% glycerol at -20°C. In this preparation the membranes are permeabilized and small molecules diffuse out, leaving the structural organization of contractile elements preserved. The preparation is now electrically inexcitable (and not alive) but contracts in the presence of ATP, generating tension comparable to living muscle. The results of these experiments were published in the MBL's journal, *The Biological Bulletin* (Szent-Györgyi, 1949). Hugh Huxley (1977) regards the glycerol extracted muscle preparation as "undoubtedly the most important and far reaching innovation introduced by Szent-Györgyi."

Having dismembered muscle completely or partially into its components, Prof characterized them biochemically and partially reconstituted them into functional units. The pressure to understand the more detailed structure of muscle led him to purchase one of the early available RCA electron microscopes, which, incidentally, he made available to summer investigators at the MBL under the supervision of Delbert Philpott. Together with Rozsa and Wyckoff, Prof undertook ultrastructural studies, and between 1949 and 1950 demonstrated that polymerized actin had a periodicity of 400Å similar to that seen in sectioned muscle and suggested that actin probably formed a continuous filament in the myofibril. They also showed that filaments were packed in a hexagonal lattice. His keen observations and his remarkable insight and intuition led to speculations (published in a booklet *Chemical Physiology of Contraction in Body and Heart Muscle*, Szent-Györgyi, 1953) that were subsequently verified by Hugh Huxley and Jean Hanson and are now currently accepted. He considered that myosin formed a hexagonal array around the actin filaments, which transmitted the tension generated by the shortening of the myosin molecules. Although little was known about the myosin molecule, he imagined it to be attached to a rigid chain. Each molecule would only shorten by about 500Å while being attached to actin during contraction, and would dissociate during relaxation. The only inaccuracy in that model is that the rigid filament is formed by a portion of the myosin (light meromyosin) itself and not by a different molecule. Hence shortening is due to movement of the heavy meromyosin, about 100Å, along the actin filament. The properties of the meromyosins were elaborated by Prof's cousin Andrew Szent-Györgyi at the Institute for Muscle Research at the MBL following earlier studies on these trypsin-generated fragments by Prof's students Gergely and Mihayli.

A number of Prof's former Hungarian students and associates joined the Institute for Muscle Research for short periods, using it as a stepping stone in their immigration to the United States. When I arrived at the Institute in 1959 as a Rockefeller Fellow on leave from South Africa, only Andrew (Csuli) Szent-Györgyi, his wife Eva, and Bob Middlebrook from England were working there on muscle.

Submolecular Biology

Prof had left the field of muscle research. In 1954, he and Jane McLaughlin, his most devoted associate who worked with him for 34 years until his death, published a paper on the quenching of fluorescence by flavinoids. He was engrossed, with George Karreman and Richard Steele, in studying excited states of amino acids and proteins. He later investigated charge transfer reactions and free radical formation in a number of substances of biological interest together with Irvin Isenberg, a fine biophysicist. Prof developed his new ideas in a booklet *Bioenergetics* (Szent-Györgyi, 1956) in which, naturally, he also discussed muscle, but I had not realized to what extent his interests had diverged. I had been stimulated by his previous booklets on muscle, particularly by his speculations on water as the "mater" of life. Indeed, I had obtained interesting results in South Africa using heavy water on a variety of muscle preparations, but had delayed publication hoping to get opinions on interpretation from the master of muscle in person. But his response to my findings was, "they are very interesting, publish them; the more I know about muscle the less I understand it." And with excitement, his blue eyes sparkling, he continued "a new dimension must be explored to understand the meaning of life," and proceeded to explain his ideas on charge transfer reactions and suggested its possible application to muscle. I naturally accepted the exciting challenge, with some trepidation, and investigated the effects of a number of electron donors and acceptors on muscle preparations. I noted to my surprise that toluidine blue stained glycerinated muscle metachromatically. For a fleeting moment Prof became engaged once again in extracting muscle. Using dye as a marker, Prof extracted a protein which he designated "metin." We pointed out in the publication, however, that the major component in the extract resembled tropomyosin and reported on the presence of a minor component which later was identified by others to be troponin.

Prof worked daily at his bench (scattered with test tube racks, pipettes, and spatulae) devising new imaginative experiments, mixing carefully selected chemicals in test tubes, and looking intently at color changes. When he formulated his ideas he imparted an exhilarating excitement and lust for discovery. His eyes sparkled when asked about his progress and with an undaunted spirit of optimism and childish confidence he would repetitively explain how his ideas were developing towards his understanding of the nature of life. His imaginative vision of molecules, atoms, electron orbitals, and energy levels was remarkable. But Szent-Györgyi, one must remember, never claimed to know quantum chemistry in any depth. Irvin Isenberg, who was the theoretical backbone, measured, with the technical assistance of Spencer Baird, free radical formation in an electron spin resonance spectrometer, one of the first manufactured in the U.S.A. Having challenged himself in 1941 to go in a direction of a new biochemistry Szent-Györgyi sailed the uncharted sea of a "Submolecular Biology," ignoring the progress in molecular biology, with a compass of his own making.

Cancer Research

In the early 1960s, Szent-Györgyi ventured into the stormy sea of cancer research, turning his rudder towards a cure for cancer. He may have been driven in this direction because both his beloved wife Marta ("Profne" as she was called in the lab) and later his daughter Nelly were afflicted by the disease. Once, in a state of anguish (and determination), he said to me,"I have a debt to settle with Nature, my ideas will lead to an understanding of cancer." He also began formulating general speculative concepts on cancer. He imagined that when life began with no oxygen in the atmosphere primordial cells grew uncontrollably. This he called the α state. Later in evolution, when oxygen appeared, cells assumed the β state, and growth was controlled by oxygen-accepting free electrons. The cancer cell, he claimed, reverted from the more developed β state to the anoxic α state. Together with Andrew Hegyeli he prepared extracts from various tissues including the thymus, which he earlier had thought might have a factor alleviating myotonia in goats. Based on preliminary observations on ascites tumors in mice, they claimed to have isolated a factor, retine, which retarded cancer growth, and speculated that retine counterbalanced a growth-promoting factor, promine. Retine was apparently also isolated from human urine, vast quantities of which Prof obtained from army personnel at the Otis Air Force Base near Woods Hole. He jokingly said that he had the American army urinating for him. Based on a hunch that retine might be a keto aldehyde, he and Laszlo Egyud then worked on methyl glyoxal, claiming that it was an anti-cancer agent. Prof was theorizing on the role of electrons in the interaction of glyoxals, particularly with amines and proteins but at the time he had no one to do ESR studies. In 1975, Ronald Pethig and Peter Gascoyne, both from Wales, resumed studies on free radicals, mainly in relation to quinones, semiquinones, and ascorbic acid interactions and cancer cells in culture. The two final manuscripts on the subject were completed shortly before Prof's death and published posthumously (Gascoyne *et al.*, 1987a, b).

During this period of *Submolecular Biology*, the title of another booklet (Szent-Györgyi,1960), Prof also expressed his views, often novel and provocative, in articles on the mechanisms of muscle contraction and the biological role of water, on teaching and expanding knowledge, on the horizons and promise of life and medical science, and on scientific and artistic creativity. When he lost his funding from NIH, he criticized publicly, again displaying courage, the requirements (which he was not able to fulfill) for writing a successful grant application. He also criticized the government for inadequately supporting science. He referred to a remark made by Szilard,"don't lie if you don't have to," and claimed that he had to lie in the grant application. "I often do not know what I am going to do the next day. I expect to think that up during the night. How could I tell then what I would do a year hence." "If one knows what one will do and find it, then it is not research any more and is not worth doing." "The NIH wants detailed projects, wants the applicants to tell exactly what they will do and find during the tenure of their grants, which excludes unexpected discoveries on which progress depends." He argued in support of his intuitive (what John Platt called the Dionysian) approach but at the same time also recognized the importance of systematic (Apollonian) investigations.

Politics Came to Him

Szent-Györgyi's dynamic and colorful personality had an impact that reached far beyond the laboratory. On his return to Szeged from Cambridge in 1930, he created an intense cultural life among students and associates surrounding him. He encouraged them to listen to music and go to the theater. He also produced *Hamlet* with student actors. He fostered a democratic approach in the University which often resulted in conflicts with administrators who had been raised in a semifeudal country. As Rector in the University, he broke down barriers between faculty and students. But the tide of fascism was rising, and as Prof put it, "politics came to him." In an attempt to save Hungary from German domination, Prof went on a secret mission to Istanbul to contact the British Secret Service. However, his activities were discovered by Nazi spies and he had to go into hiding to escape the Gestapo. He nevertheless joined an underground Resistance Front whose aim was to prevent the siege of Budapest. As told by his close friend and colleague Zoltan Bay, the plan was to establish a joint strategy between the advancing Russian army and an anti-Nazi segment of the Hungarian army. Prof had the risky mission of flying over to the Russian front to arrange the plan and to communicate the details by radio to Zoltan Bay in his research laboratory. Two days before Prof was to fly across, the entire Resistance Movement was betrayed by Hungarian spies. Participants were arrested, but Prof managed to obtain safe hiding in the Swedish Embassy. During these unbelievably difficult times amidst chaos and turmoil, the important research on muscle continued in his laboratory. As already mentioned, the manuscript on this work was sent to Stockholm for publication to ensure that the results and conclusions were not lost should Prof be killed. Prof had sent the manuscript to his friend Hugo Theorell in Stockholm, and apparently this correspondence enabled the Nazis to track Prof down in the Swedish Embassy. Fortunately the Embassy staff were alerted and arranged for Prof's escape just in time.

When the Russian troops marched into Hungary there was an initial ray of hope of liberation, but the behavior of the Russian army of occupation was outrageous. They closed off roads and herded masses of people into camps with unsanitary conditions; those who survived were sent off in sealed trains to unknown destinations. No one realized at the time that they were enslaved in Russian labor camps. Because he and his family had, for safety purposes, been escorted, at the order of Molotov, to Moscow for a brief period when the Russians pushed out the Germans, Prof thought he would return there and have access to Stalin to inform him of the state of affairs in Hungary. But a lower official disregarded him and sent him home. The writing was on the wall. Nevertheless Prof had hopes for improvement.

In 1945 Prof was appointed Head of The Institute of Medical Chemistry in Budapest. He immediately arranged a free kitchen, run by his wife Marta, to feed his laboratory associates. He also arranged for the distribution of food supplies to members of the Hungarian Academy of Science. Food was in short supply during the post-war period, and he had to use his influence with the Russian authorities and friends to keep his community alive. Laszlo Lorand recalls how appreciative he was to receive food not only for himself but also for his widowed mother. All needs were often difficult to meet. As Andrew Szent-Györgyi remembers, von Bekesy (who later, while at Harvard, received the Nobel prize for his work on hearing) always wanted a large supply of food, enough to feed his neighbors because he was unable to eat while they starved.

Prof also tried to democratize education by helping to organize a National Council of Public Education. He tried to introduce progressive procedures in the Academy of Sciences but his efforts were met with strong conservative opposition. Undaunted, he founded a new Academy of Natural Sciences. Later he assisted the government in fusing the two Academies; he declined an invitation to become the new Academy's President. Together with his friend Zoltan Kodaly, the great composer, he assisted in revitalizing the cultural life in a period of depression and despair. He hated the military and, according to Andrew Szent-Györgyi and Laszlo Lorand, he led a movement to convert the Military Academy that adjoined the Medical School into student dormitories. Szent-Györgyi, the first Hungarian Nobelist and held in great esteem, was made an honorary member of the Hungarian parliament. He promptly moved a motion to abolish the Armed Forces and for Hungary to come under the protection of the United Nations. His derogatory comments about the army angered the Secretary of Defense, who challenged him to a duel. Prof's discretion on this occasion was the better part of valor.

As usual, Prof created an atmosphere of fun and excitement in his lab. Always full of humor, he arranged parties and dances for his associates and students. He insisted on a game of volleyball after lunch in the courtyard and, as Lorand put it, one had to first qualify as a volleyball player before joining Prof's laboratory.

While in Switzerland for a skiing vacation in 1946, Prof learned that a close friend, a banker who helped finance the laboratory, was imprisoned. Prof immediately sent telegrams to the Hungarian Prime Minister threatening to bring the arrest to the attention of the outside world. His gamble that the local Communists would find this embarrassing worked. His friend was released and soon joined him in Switzerland. Prof had been very disenchanted with the Communist regime in Hungary and the ruthless domination of Hungary by the Soviet Union. When he learned of the barbaric treatment of his friend he decided not to return to his homeland. In 1947 Prof found a haven at the MBL. Through his international contacts and influence he also helped many friends and associates to get out of Hungary by arranging for invitations to be extended to them by a number of English and European investigators. Others proceeded to the United States. Among them were Lorand, Mihalyi, Gergely, Varga, Rosza, Erdos, Csapo, Lajtha, and Andrew and Eva Szent-Györgyi. Some who arrived in the States joined him but had to travel twice a week for visa purposes to New York to teach at the New School. Money was short and times were difficult. Even though he opposed Communism in Hungary, Prof, being an immigrant from a Communist country, was under surveillance by the F.B.I. As everyone knows, Prof had to raise funds from various sources to pay his MBL laboratory rental fees and to cover other expenses. Prof was privileged to be given the first and, for a long time, the only all-year-round laboratory at the MBL.

Having fought Hitler and Stalin, Prof continued to be concerned with the serious problems facing mankind and was outspoken against the involvement of the U.S.A. in the Vietnam War. In a little book, *The Crazy Ape* (Szent-Györgyi, 1970), he saw hope in the youth of the world and called upon them to organize and exercise their democratic power to create a new world. He unceasingly voiced his forceful opposition to the madness of the nuclear arms race. He publicly debated against Edward Teller and joined Leo Sziilard in his movement of scientists for a peaceful world. He dreamed of science opening up "endless possibilities for expansion if we work together instead of snatching small advantages from one another." He hoped science would help us "understand and master ourselves" enabling us to create wealth and beauty "which cannot be pictured today by the keenest imagination." Hans Krebs, on acknowledging his personal gratitude to Szent-Györgyi for having influenced him to go to Cambridge and to get out of Germany during the Hitler period writes, "Another reason for my special affection for Albert Szent-Györgyi, which I am sure I share with many, is my close affinity to his outlook on life and my admiration for his courage and eloquence in expressing his philosophy. In the *Bulletin of the Atomic Scientists* he published what he called "A Little Catechism": 21 succinct paragraphs of his passionate concern for real peace and the welfare of fellow man, his moral courage which gives him strength to speak out in in support of unpopular truths, and his profound wisdom in discerning the essentials of life."(see Szent-Györgyi, 1975).

Woods Hole: Science and Fun

Prof led an intense intellectual life which he enjoyed; he had fun in his laboratory. At the MBL he was not disturbed by the usual institutional responsibilities, and Homer Smith, the general manager, attended to his needs with warmth and friendship. For recreation, he played chess with Csuli and he very much enjoyed vigorous physical exercise: table tennis, tennis, volleyball, skiing, horseback riding, and swimming. At the age of 80 he started water-skiing. He would enjoy adventurous swims around Penzance Point, boating and fishing; Tay Hayashi showed him some of the good spots for striped bass soon after he arrived in Woods Hole. Prof always used a big hook and when asked why, he said "I find it more exciting not to catch a big fish than not to catch a small one." Using the same philosophy in the lab he made some big catches but also lost many. His successes and failures were never a hindrance to him. He would pronounce "I never look back, I only look forward." And he never feared making a mistake. He wrote " There is but one safe way to avoid mistakes: to do nothing or at least to avoid doing something new."

Prof played an important role in the life of the summer scientific community at the MBL. He taught in the physiology course and established a section on muscle. His lectures, animated and exciting, stimulated generations of students and investigators. He continued to give an annual lecture, fascinating the audience with demonstrations of some new color reaction in gigantic test tubes. He enhanced the intellectual life in Woods Hole. Numerous scientists of renown passed through to visit Prof in the summer. Among them were Michaelis, Warburg, Meyerhof, Loewi, Cori, Ochoa, Huggins, Pauling, Bernal, and Dirac. He established a rapport with summer "regulars" such as George Wald (an outspoken mutual admiration developed between them), Burr Steinbach, Stephen Kuffler, Hans Stetten, Rachmiel Levine, Tay Hayashi, Hugh Huxley, Annemarie Weber, and Denis Robinson, to name but a few.

In the summer of 1959 Prof organized in his home (The Seven Winds, at Penzance Point) a private conference aimed at building bridges between quantum electronic phenomena and biochemistry and biology. Among the participants were Hugo Theorell, Theodor Forster, John Platt, William Arnold, Zoltan Bay, Alberte and Bernard Pullman, William McElroy, Irvin Isenberg, Gregoria Weber, Sydney Vellick, Richard Steele, Irving Klotz, Henry Linshitz, Richard Bersohn, and Michael Kasha. Here were gathered a mix of biochemists, molecular spectroscopists, physicists and biophysicists, photochemists, and quantum chemists. Szent-Györgyi led the knocking of heads to gain a better understanding of the meaning of life. Kasha writes in a personal communication, "The conference was one of the major educational and stimulatory events in the lives of the participants. Most of us had not met before and these meetings not only had a profound effect on the direction of our research interests, but led also to new lifetime friendships." This conference was perhaps the seed from which developed, in 1968, the organization of annual conferences in Quantum Biology held in Sanibel, Florida. Here Szent-Györgyi again played a prominent role. Kasha gives examples of how Prof's experiments and observations "stimulated others to embark on wholly new directions of research." He cites examples of Prof's experiments based on his notion of the triplet states of molecules. Prof illustrated to him how the green fluorescence of acridine orange in water at room temperature changed to a red phosphoresence when frozen at 77K, and ascribed it to crystalline water causing a spin orbital perturbation by enhancing the triplet state. Kasha and his graduate student were challenged and eventually determined that there was no enhanced spin orbital coupling and that the frozen water simply permitted dimerization of the dye molecules. He says, "This contact with Szent-Györgyi launched my laboratory and me into a decade of research on molecular exciton effects." The same finding of Szent-Györgyi published in *Submolecular Biology* led Beukers and Behrend to freeze thymine in water and to then irradiate it with UV light thus producing dimers. This in turn led to much interest in photodimerization of thymine in DNA photoinactivation and repair mechanisms. Szent-Györgyi's ideas, however faulty in formulation, bore fruit in their impact on investigators of quantum biochemistry. Many "solid scientists," however, strongly disapproved of his flights of imagination and hasty pronouncements. But who can tell where these pioneering inroads at the electron level will lead in our future understanding of the living process? Was Prof offbeat, or ahead of his time?

Full of the joys of life, Prof would invite members of his Institute and other friends to fancy dress parties during the cold winter months, and his and Marta's costumes were often the most elegant and imaginative. A particularly well-known and hilarious party was held at Prof's home during the summer of 1954; members of the RNA Tie Club attended. The Club, organized by Gamow, contained 20 members: pioneers in molecular biology, each representing an amino acid (and Szent-Györgyi was invited as an honorary member). During this summer, Gamow was a guest in Szent-Györgyi's cottage and Crick, Brenner, and Watson spent time there speculating on the genetic code.

In the summers, Prof participated in the social activities of the MBL community, attending the music concerts on Sunday evenings and, until about the age of 70, joining the Saturday evening parties in the MBL Club where he mixed with the young investigators and students who enjoyed his boundless anecdotes and jokes. How thrilling and stimulating for the younger generation! On many occasions, Prof led the traditional Fourth of July parade down the main street of Woods Hole.

Prof found a haven at the MBL, and he could not have functioned as he did at any other institution.

Literature Cited

Bailey, K. 1947. Chemical basis of muscle contraction. *Nature* **160**: 550-551.

Gascoyne, P. R. C., R. Pethig, and A. Szent-Györgyi. 1987a. Electron spin resonance studies of the interaction of oxido-reductases with 2,6-dimethoxy-p-quinone and semiquinone. *Biochim. Biophys. Acta* **923:** 257-262.

Gascoyne, P. R. C., J. A. McLaughlin, A. Szent-Györgyi, and R. Pethig. 1987b. Free radical investigations as a tool for the study of quinone detoxification by whole cells and enzymes. *Chem. Scr.* **27A**: 125-129.

Hammerstein, E. 1937. *Presentation Speech in Nobel Lectures: Physiology and Medicine 1922-1941.* Elsevier Publishing Co., New York. Pp. 435-439.

Huxley, H. 1977. Past and present studies on the interaction of actin and myosin. Pp. 63-74 in *Search and Discovery: A Tribute to Albert Szent-Györgyi,* Benjamin Kaminer, ed. Academic Press, New York.

King, C. G., and W. A. Waugh. 1932. The chemical nature of Vitamin C. *Science* **75**: 357-358.

Krebs, H. 1977. Pp. 3-15 in *Search and Discovery: A Tribute to Albert Szent-Györgyi*, Benjamin Kaminer, ed. Academic Press, New York.

Perutz, M. 1988. *The New York Review of Books*, October, Vol. **13**: 21-23.

Svirbely, J. L., and A. Szent-Györgyi. 1932. Hexuronic acid as the antiscorbutic factor. *Nature* **129**: 576.

Szent-Györgyi, A. 1935. Uber die Bedeutung der Fumarsaure für die tierische Gewebsatmung . Einleitung, Ubersicht, Methoden. *Hoppe-Seyler's Z. Physiol. Chem.* **236**: 1-20.

Szent-Györgyi, A. 1936. Uber die Bedeutung der Fumarsaure für die tierische Gewebsatmung. Einleitung, Ubersicht, Methoden. *Hoppe-Seyler's Z. Physiol. Chem.* **244**: 105-116.

Szent-Györgyi, A. 1940. On protoplasmic structure and functions. *Enzymol. Acta Biocatalytica* **9**: 98-110.

Szent-Györgyi, A. 1945. Studies on muscle from the Institute of Medical Chemistry, University of Szeged, 1944. *Acta Physiol. Scand.* **9**, Suppl. XXV.

Szent-Györgyi, A. 1949. Free-energy relations and contraction of actomyosin. *Biol. Bull.* **96**: 140-161.

Szent-Györgyi, A. 1953. *Chemical Physiology of Contraction in Body and Heart Muscle.* Academic Press, New York. 135 pp.

Szent-Györgyi, A. 1957. *Bioenergetics.* Academic Press, New York.

Szent-Györgyi, A. 1960. *Introduction to a Submolecular Biology.* Academic Press, New York.

Szent-Györgyi, A. 1963. Lost in the Twentieth Century. *Ann. Rev. Biochem.* **32**: 1-14. (Reproduced in *The Biological Bulletin*, 1988. **174**: 191-240).

Szent-Györgyi, A. 1970. *The Crazy Ape*. Philosophical Library, New York.

Szent-Györgyi, A. 1975. A little catechism. *Bull. AT. Sci.* **31**: 5.

Keith R. Porter
(1912 -   )

Porter and the Fine Architecture of Living Cells

Introduction by Robert Goldman

Keith Robert Porter, D.Sc., Ph.D., L.L.D., was born on June 11, 1912, in Yarmouth, Nova Scotia. He spent his formative years there and received his B.Sc. in Biology from Acadia University in 1934. He completed his Ph.D. in Biology at Harvard in 1938 and did postdoctoral research with Professor Gerhard Fankhauser at Princeton until the outbreak of World War II in 1939. In that same year he moved to the Rockefeller Institute for Medical Research in New York City, initially to work with Dr. James Murphy and ultimately to become the Chief of the Laboratory of Cytology. Drs. George Palade and Philip Siekevitz were among the members of this research group.

Keith's first project at the Rockefeller involved frustrating attempts to grow and transform cells in culture, an experience that led him eventually to organize the Tissue Culture Commission which, in turn, became the Tissue Culture Association. During this period he spent many hours observing live cells in the light microscope. He quickly realized that there was much more to be seen than met the eye. This early realization and his insatiable curiosity about cell structure were ultimately to lead to his interest in electron microscopy. By the mid-1940s, electron microscopes (E.M.) became commercially available, but biologists had very little access to them. Keith immediately realized the potential value of the higher resolution afforded by this new tool, for delving into the substructure of cells. Through Albert Claude, another of his colleagues at Rockefeller, Porter met Ernest Fullam, then the electron microscopist at the Interchemical Corporation in Manhattan. Fullam gave him the basic conditions necessary for seeing objects in the E.M., the most important of which was that they had to be extremely thin. Keith immediately thought of his cultured cells, which had the property of spreading into very thin sheets especially at their margins. He worked quickly to develop some important basic techniques (still in wide use in cell biology labs worldwide), such as the placement of cells on metal grids supported by a thin plastic film and the preservation of their cytoarchitecture with vapors of osmium tetroxide. It was an exciting time, as everything seemed to work easily, and soon he was prepared to inspect cultured cells with the company's E.M. He went to Interchemical one night after closing hours and with Fullam and Claude, began the first observations of animal cells on the luminescent screen of the E.M. They stayed up almost until daybreak taking sufficient micrographs for the landmark publication which appeared in the *Journal of Experimental Medicine* in 1945 (Porter *et al.*, **81**: 233-247).

Keith Porter was the first serious explorer in the amazing new territory of subcellular ultrastructure, and, as a result, he was the first to describe the fine structure of almost all of the major organelles. For example, he described the basophilic regions of the cytoplasm as the lacelike reticulum and later as the endoplasmic reticulum. After a few years of studying whole mounts, he realized that he had to develop methods for viewing thinner sections of cells in order to define all of the elements of cytoplasmic and nuclear fine structure. As a result, he collaborated with Joseph Blum of the Rockefeller in the design and construction of the Porter-Blum ultramicrotome, which permitted the routine preparation of ultrathin sections of cells and tissues. This machine helped to open up a whole new world of tissue and cell structure.

In 1961, Keith moved to the Harvard Biology Department where he served as Chairman for two years. During his Harvard period he and his students defined in great detail the cytoplasmic microtubule system in a variety of cells, and initiated a period of tremendous excitement in studies of cell shape and motility that has lasted to this day. It was at the time of his initial introduction of the cytoplasmic microtubule and his futuristic predictions about its functions, that I first met Keith. I was a graduate student at Princeton and chaired the student committee charged with the task of inviting a single speaker for the 1964-65 academic year. I invited Keith, and he delivered a very exciting lecture that thrilled all of the graduate students. However several of the more physiologically and biochemically oriented professors warned that the microtubule was undoubtedly an artifact of both electron microscopy and Keith Porter's imagination. How wrong they were!

Tiring of Harvard by the late 1960s, and desiring a new adventure, he went west to the University of Colorado at Boulder to establish the Department of Molecular, Cellular and Developmental Biology (MCDB). This involved the construction of a new building (eventually named after him) with a large, built-in two story room to house his newest acquisition, a one million volt electron microscope. His own words best describe his feelings about this new machine: "So the vastness of the great west attracted me as it has many others. I had the opportunity there to build a new building and a new department and incorporate in it the 22 tons of metal and foundations required for this new venture." For years, Porter had wanted to go back to his whole mount preparations, first viewed in 1945, to look through thicker regions of cytoplasm. The purpose was to study in detail the so-called ground plasm that surrounded all of the cell organelles and that was thought to be the structureless cytosolic compartment. His latest hypothesis is that this is not the case and he has suggested, from his micrographs, that this region contains an extensive array of anastomosing filaments of varying thickness that interconnect all of the organelles and the various cytoskeletal systems. He now believes that this so-called microtrabecular lattice is involved in coordinating the activities of all organelles and cytoskeletal elements; that it is the structural basis of organization. This is a very controversial issue and many cell biologists don't accept the idea. However, Porter's imagination has paid off many times in the past, and this most recent theory of his may be no exception.

Following his retirement as Distinguished Professor from the University of Colorado, he decided in 1984 to take on yet another challenge back on the east coast, where he became the Wilson Elkins Distinguished Professor, and Chairman of the Department of Biological Sciences at the University of Maryland. In 1988, he retired once again, only to accept, at the age of 76, another position as Research Professor at the University of Pennsylvania where he is currently actively engaged in efforts to explore the reality and function of his microtrabecular system.

It is evident from this brief history that Keith Robert Porter is a remarkable man, and it is no wonder that he is frequently referred to as the "Father of Modern Cell Biology."

Shinya Inoué received his B.A. in Zoology from the University of Tokyo in 1944 and a Ph.D. in Biology from Princeton in 1951. He has been an Instructor in Anatomy at the University of Washington; Assistant Professor at Tokyo Metropolitan University; Assistant and Associate Professor of Biology at Rochester; the John LaPorte Given Professor and Chair of the Departments of Cytology and Anatomy at Dartmouth; and Director of the Program in Biophysical Cytology at the University of Pennsylvania where he continues as Adjunct Professor. Since 1986 he has held the position of Distinguished Scientist at the Marine Biological Laboratory.

During his long association with the MBL, he has served as an Instructor in the Physiology Course, Instructor-in-Chief of the Analytical and Quantitative Light Microscopy Course and has been a member of the Board of Trustees. His awards include a Lalor Foundation Fellowship from the MBL, an Honorary Fellowship from the Royal Microscopical Society, and the Lewis S. Rosensteil Award for Distinguished Work in Basic Medical Research. He is a Fellow of the American Association for the Advancement of Science and the American Academy of Arts and Sciences.

Ever since his student days, he has been attempting to elucidate the molecular architecture of the cytoplasm and the nucleoplasm, with special emphasis on the mechanisms of mitosis. In pursuit of these interests, he has become a world-renowned innovator in the area of light optics, especially as applied to living cells. As a student under Professor Katsuma Dan, he began defining the elements of the mitotic apparatus *in vivo* through the use of polarized light optics. His first encounter with mitotic spindles took place under unusual circumstances; it was during an air raid blackout in the home of Professor Dan while he was an undergraduate at the University of Tokyo. In the total darkness afforded by the air raid, Shinya began to work out the conditions for the direct observation of the birefringent structures comprising the mitotic spindle in dividing sea urchin eggs.

Soon after World War II ended, Shinya returned with Professor Dan to the Misaki Marine Biological Station, which had been occupied for some time by American troops. It was during this period of his career that he worked day and night in attempts to directly visualize the birefringent fibers of the mitotic apparatus. At that time, many cytologists felt that spindle fibers were an artifact of the fixation methods used to visualize them. He spent many hours each day in total darkness, peering down the eyepieces of a microscope that he had assembled from disparate parts, using the intense light of a high pressure mercury lamp. Finally, through his constant tinkering with the optical system, he was able to distinguish spindle birefringence *in vivo*. However, he was not happy with this achievement because he could not resolve individual spindle fibers running between the chromosomes and the poles. A breakthrough occurred while he was attempting to improve the extinction factor of his polarization microscope. Quite remarkably the birefringence of the mitotic spindle disappeared while he was making a simple modification of the optical system. Professor Dan told him that if he had not been so insistent on tinkering with the microscope, he would still be able to see the spindles. As it turned out, the chance observation led to the development of the rectified lens system for polarized light. The use of this improved polarization microscope led to the first convincing description of spindle fibers *in vivo* in a paper that appeared in 1953 (Inoué, *Chromosoma* **5:** 487-500). This paper settled the battle over the reality of spindle fibers, which had been raging among cytologists for over 50 years. During the next few years, Shinya published numerous papers that described, for the first time, the dynamic nature of the mitotic spindle using drugs, temperature, and changes in hydrostatic pressure. From these observations he developed the Dynamic Equilibrium Model for spindle function and chromosome movements. This model still stands as the basis for all of the most recent ideas regarding mitotic mechanisms.

It is most appropriate that Dr. Inoué has been asked to lecture in honor of Keith Porter, as it turns out that the major component of the mitotic spindle fiber is the microtubule; one of those cytoplasmic structures first brought to the world's attention by Porter and his students.

Keith Porter and the Fine Architecture of Living Cells

Shinya Inoué

Marine Biological Laboratory

WE HONOR DR. KEITH PORTER, a modern pioneer in cell biology and a prominent member of the MBL who is still actively pursuing research in the laboratory today. Keith Porter and I both use microscopes to explore cell fine structure and behavior, although our approaches are somewhat different. Keith has emphasized the high resolution of the electron microscope, while I have pushed the development and use of the light microscope. The electron microscope gains its high resolution by virtue of the short wavelengths of accelerated electrons; but to gain this advantage, the electron beam and specimen must be placed in a high vacuum. The light microscope has a limited resolving power owing to the much longer wavelength of light, but we do have the advantage of being able to observe the cells directly in their living state. Regardless of these differences, we must keep in mind that the more we magnify the microscopic world, the more limited becomes our field of view at any one instance. Hence the admonition of Daumier regarding his contemporary, Monsieur Babinet, applies to both Keith and myself (Fig. 1).

To be aware of such limitations, and still plunge ahead and explore the detailed architecture of life, one seems to require certain character traits. Keith certainly has that. One has to have perseverance; we both seem to be known for our stubbornness. And finally, one has to be prepared to have many eyes cross check our observations. Keith and I have both been blessed with a number of excellent students, trainees, and collaborators, although Keith Porter has had a much greater group of followers, as he richly deserves.

Figure 1. *Monsieur Babinet being apprised of a visiting comet by his maid; by H. Daumier. Author's comment: "Yes, Monsieur Daumier, we may indeed miss major happenings by concentrating on microscopic observation. But then, without intent inspection and considered deductions, how could we have perceived the spectacular universe, our wondrous earth, and the miraculous inner world? Perhaps the telescope and microscope do whisper Nature's elegant secrets and illuminate hopeful paths for those who will listen." (The figure, which is M4182, DEL 3080, M. Babinet prevenu par sa portiere de la visite de la comete, Honore Daumier, French, 1808-1879, Lithograph, Babcock Bequest, is reprinted here courtesy of the Museum of Fine Arts, Boston.)*

Despite his shining fame and success in electron microscopy, Keith did not always work with fixed cells or with the electron microscope. In fact, using enucleated frog eggs and heterospecific sperm, he was one of the first to look into the mutual influences of the cytoplasm and the cell nucleus in embryonic development (Porter, 1942). In his early efforts, Porter worked extensively with living tissue culture cells, and was actually the organizer and first president of the Tissue Culture Association in 1946. He was re-elected as president of the Association three decades later. He was also an organizer of, and, in 1961, the first president of the American Society for Cell Biology; he was re-elected to that post 16 years later.

Although Keith worked directly with living cells during his earlier career, he succumbed to the temptations of the higher resolution promised by the electron microscope almost as soon as it became available to biologists. Tutored initially by Albert Claude, Keith entered a prolific, pioneering examination of cell fine structure using the electron microscope. Figure 2 is an early transmission electron micrograph of the cytoplasm captured by Porter. He selected very thin regions of tissue cultured cells so that the electron beam could penetrate the region. The cells were treated with osmium vapor to quickly fix visible particles in place, and to provide good contrast for the electron optical image. With these early micrographs, Porter called attention to the fine, granular cytoplasmic inclusions which, he deduced with foresight, were active in protein synthesis (Porter and Kallman, 1952). He also noted the various intracellular membranes and filaments that appeared to play important roles in organizing the basic architecture of the living cytoplasm.

For studies of muscle fibrils, Porter initially shadowcast the specimen with heavy metal (chrome) to display details of their surface structure and fine filaments protruding from the ends (Ashley *et al.*, 1951). But on realizing the need to get better electron transmission with much less scattering of electrons, he devised the Porter-Blum ultramicrotome. With it, one could slice tissues and cells embedded in plastic into sections so thin that the thickness of a human hair could be sliced into a thousand or more thin sheets. The result was striking. In a figure of a thin section of muscle, published by Franzini-Armstrong and Porter (1964), we can see not only the thicker myosin filaments and the thinner actin filaments which interdigitate and slide against each other to make muscle contract (as shown by Huxley and Hanson, 1954; Huxley and Niedegerke, 1954), but also we can clearly see the membrane components which couple the neural signal with the contraction of muscle (as demonstrated by Huxley and Taylor, 1958).

Porter explored one type of cell after another with this powerful tool, first with the transmission electron microscope, then scanning electron microscope, and finally with the high voltage electron microscope which allows greater electron penetration. But when it

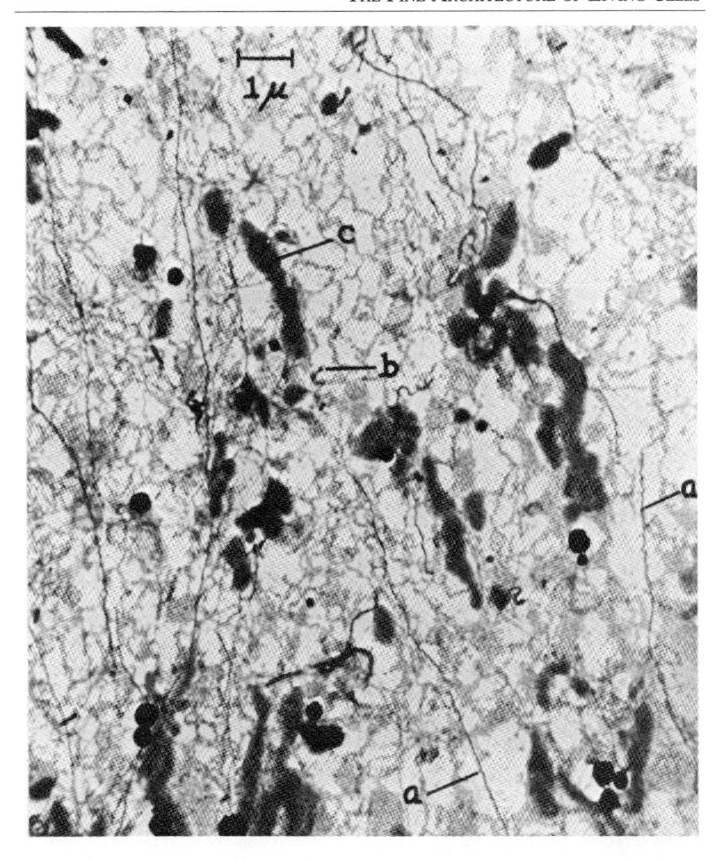

Figure 2. *Early electron micrograph of cytoplasm. Porter chose a very thin region of an osmium-fixed tissue cultured cell to show the widespread occurrence of characteristic fine particles and filaments in the cytoplasm. He had the foresight to deduce that the particles were engaged in protein synthesis. (From Porter and Kallman, 1952. Photograph courtesy of Dr. Keith Porter.)*

came to dividing cells, even Keith was not infallible. His paper in the early 1960s states that the region of the cell which others thought contained spindle fibers, in fact, contained elongated membrane but not fibrils.

Ten years before that I had shown in polarized light microscopy that the spindle fibers did, in fact, exist in living cells as positively birefringent bundles of submicroscopic fibrils (Inoué, 1953). To be sure, these fibrils were labile: they appeared and disappeared depending on division stage, and would disappear reversibly by treatment with drugs such as colchicine—a cell-division inhibitor—or with low temperature, high pressure, etc. (Inoué, 1952a,b; Inoué *et al.*, 1975). Using time-lapse movies, I have shown the reversible assembly-disassembly of microtubules in living, dividing cells, and as affected by cold, high hydrostatic pressure, etc. (Inoué, 1964). These movies were taken by myself and my students with the sensitive polarizing microscope and accessories that we developed.

Before I had captured those scenes with polarized light microscopy, and before Dan Mazia and Katsuma Dan figured out how to isolate the mitotic apparatus as a coherent physical body (Mazia and Dan, 1952), many physiologists had strong arguments about why these fibers, which were generally visible only after fixation, were artifacts of fixation and not present in the living cells. In fact, that notion was so strong that when I first showed the time-lapse polarized light movies of the dividing cells in the Lillie auditorium in the early 1950s, Dr. Ethel Browne Harvey asked from the audience whether the cells were actually alive!

I had thought that by the mid 1950s I had convinced people that spindle fibers and their fibrils were not artifacts, and that they clearly were present and played important roles in mitotic cell division (Inoué, 1953; Schrader, 1953). But apparently Keith Porter was not a ready convert, and he would not buy the argument that membranes would show a different kind of birefringence, or that osmium fixation could induce the loss of some important cell structures.

That was until a better fixation was discovered by Sabatini *et al.* (1963). Very quickly Keith found that the missing filaments—microtubules—were present in profusion in all kinds of cells playing many important roles, including those in the mitotic spindle and neuronal processes (Fig. 3). With his famous paper at the Ciba Foundation Symposium (1966; see also Tilney and Porter, 1965), Keith became "Mr. Microtubule." But to show that the microtubules

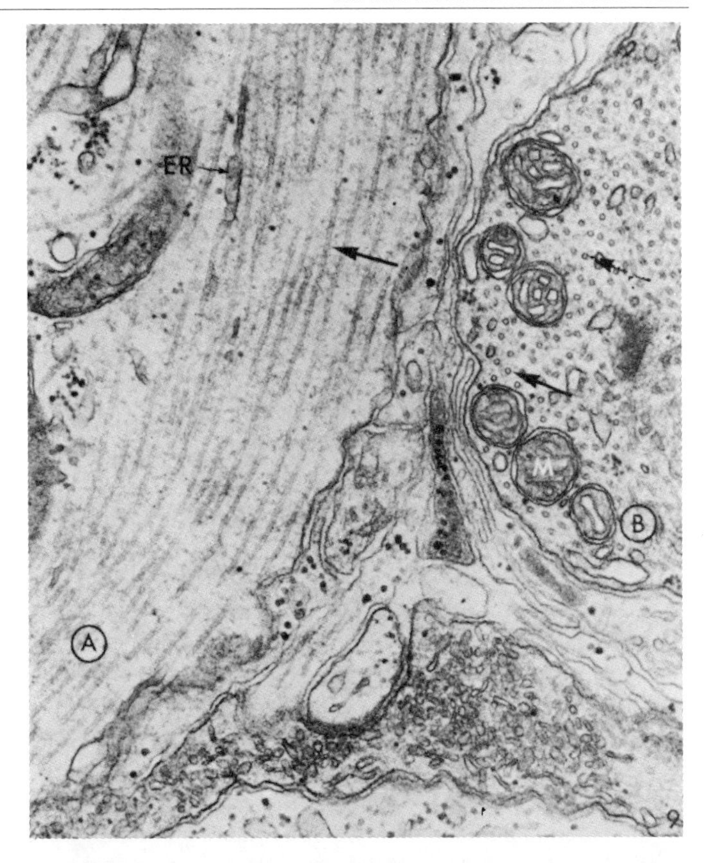

Figure 3. *Thin section of frog brain. In this electron micrograph Porter shows the longitudinal- (A) and cross- (B) sections of the microtubules (long arrows) which populate the Purkinje cell dendrites. (From Porter, 1966. Photograph courtesy of Dr. Keith Porter.)*

were present in the spindle in living cells, he came to borrow a slide of mine showing the birefringence of spindle fibers in the living cell and proving that the new electron microscope image was not an artifact!

So much for leg pulling and reminiscing. Our Centennial celebration is over, and, in fact, as of July 17, 1988, we entered the Second Century of the MBL. I felt that this lecture was an appropriate occasion to share with you some ongoing studies with the light microscope and about some future hopes for light microscopy.

For the past ten years, I and my colleagues, and Bob and Nina Allen and their associates, have applied video or television technology to the microscope, vastly improving what we can see with the light microscope. By using appropriate video equipment, we were able to see objects far below the previous limit of resolution, obtain images with much improved quality, and do this much faster than hitherto was possible (Allen *et al.*, 1981a, b; Inoué, 1981, 1986, 1987; Allen, 1985). For example, Figure 4 shows the improvement of image contrast and detail that can be brought about by the use of video. Video tapes now can show details in cellular structures and activities that could not be captured without the use of video. The birefringence of beating axostyles in, and flagella on, hypermastigote protozoa can be seen clearly in video-enhanced polarized light microscope images, as can the rotating flagella on a bacterium. With differential interference contrast microscopy, we can also see the striking extension of the sea cucumber sperm's acrosomal process (Tilney and Inoué, 1982).

Instrumentation video cameras that became available in the late 1970s help the microscopist by raising contrast, increasing the speed of capture, and making it possible to see objects way below the limit of resolution. Some video cameras also help with low light imaging. These improvements, which are brought about by adjusting the analog video signals, are made even more striking when the video signal is processed through a digital image processor (Inoué, 1986, 1987). For example, Joe and Jean Sanger have microinjected, into living tissue culture cells, selected proteins such as actin (which makes up the core of thin filaments of muscle) and light chain components of myosin (found in the thick filaments of muscle), which were conjugated to fluorescent dyes that do not disturb the protein function (Mittal *et al.*, 1987). The incorporation of the protein in the cell is signaled by the fluorescence that becomes localized in specific regions of the cell.

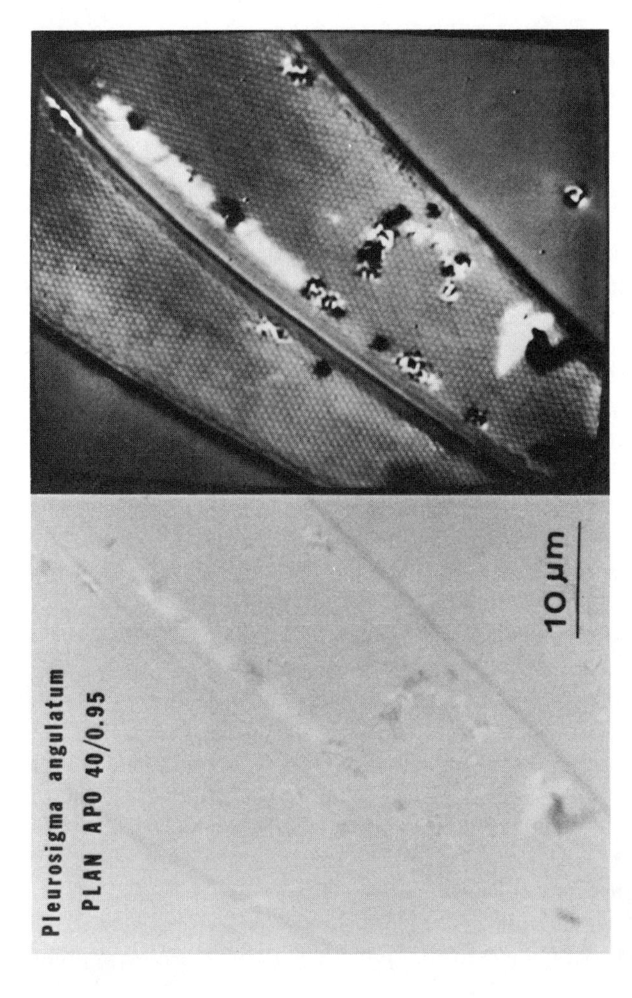

Figure 4. *Diatom in a polarizing microscope, viewed with and without video enhancement. Panel A is before, and panel B after, analog enhancement of the low-contrast image with a Newvicon video camera equipped with automatic black level and gain controls. (From Inoué, 1981.)*

Ordinarily, a cell injected with such dyes becomes sensitive and cannot tolerate the intense, short wavelength light used to excite the fluorescence. The dye itself also tends to bleach in the intense, exciting light. The Sangers used a sensitive SIT (silicon intensifier target) video camera, which allowed the exciting light level to be cut way down. Then they processed the weak video signal with a digital image processor (Image-I/AT Processor made by my son Ted's company, Universal Imaging Corporation of West Chester, Pennsylvania). This way they reduce the noise inherent in low light level imaging, the signal is enhanced to give these striking time-lapse images, and the cell remains happy so that the same individual can be observed for hours.

In one video scene, a tissue cultured cell is stained with Di-IO, a fluorescent dye that stains the membranes inside the cell. We can see how dynamically the cell's membrane system is changing all the time, a point made by Keith many years ago, but now clearly visible in living cells. Another video scene shows a dividing fibroblast cell to which fluorescent actin molecules are injected. In interphase, the actin is incorporated into the stress fibers. But the stress fibers fall apart and the fluorescence disperses, as the cells prepare for mitosis. Near the end of mitosis, the actin molecules form a belt in the cell cortex between the separating chromosomes, and cleave the cell into two between the daughter nuclei. A very similar behavior is seen for fluorescently labeled myosin.

These scenes unequivocably establish the presence of an acto-myosin contractile ring in cell cleavage and also show the assembly-disassembly properties of actin and myosin filaments.

The chromosomes themselves are separated by the spindle fibers made up primarily of microtubules. As I discussed earlier, the microtubules are birefringent, *i.e.*, they show direction-dependent properties to polarized light so that they appear brighter or darker than the background between crossed polarizers in the presence of a compensator (Inoué, 1986, 1988). With digital enhancement and with further improvement of the microscope optics, we now get images of microtubules in mitosis that are far better than we were able to gain just a few years ago. In fact, we can now visualize individual microtubules (which measure only 25 nm, or 1/4000th the diameter of human hair) directly with the light microscope (Figs. 5, 6). This does not mean that the resolution of the light microscope has been raised

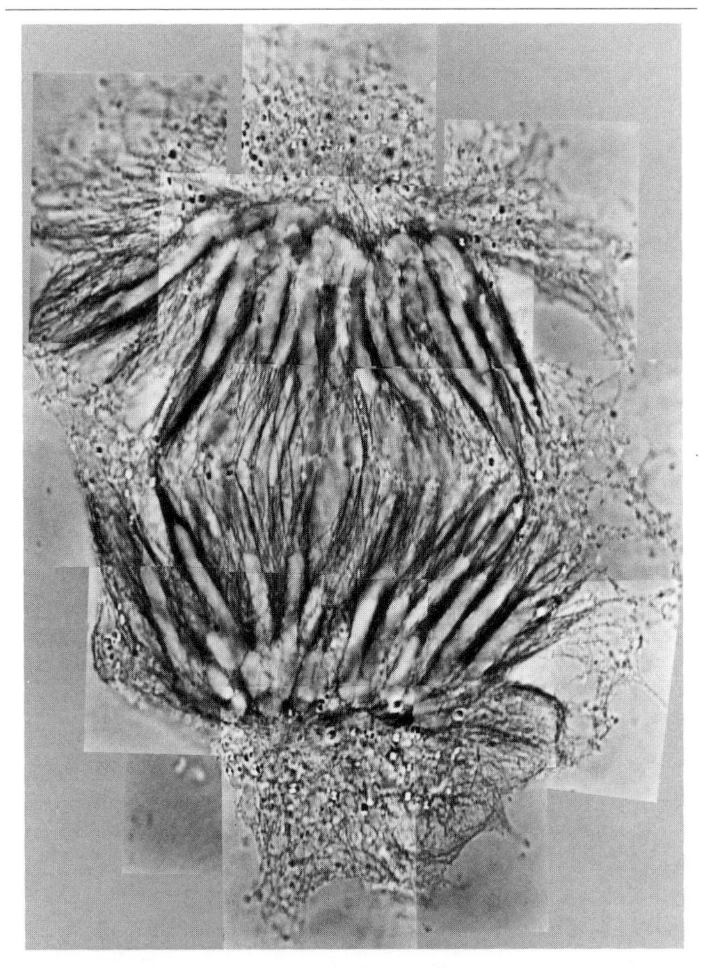

Figure 5. *Microtubules of plant cell in late anaphase. The* Haemanthus *endosperm cell was fixed as a whole mount, and stained with 5-nm diameter colloidal gold conjugated to antitubulin, by Andrew and Wishia Bajer. Individual microtubules, only 25 nm in diameter, stand out clearly as thin black threads between the large, white, late-anaphase chromosomes. This montage was generated from thin optical sections captured with the author's video microscope and processed with the Image-1/AT digital image processor. (From Inoué, 1987.)*

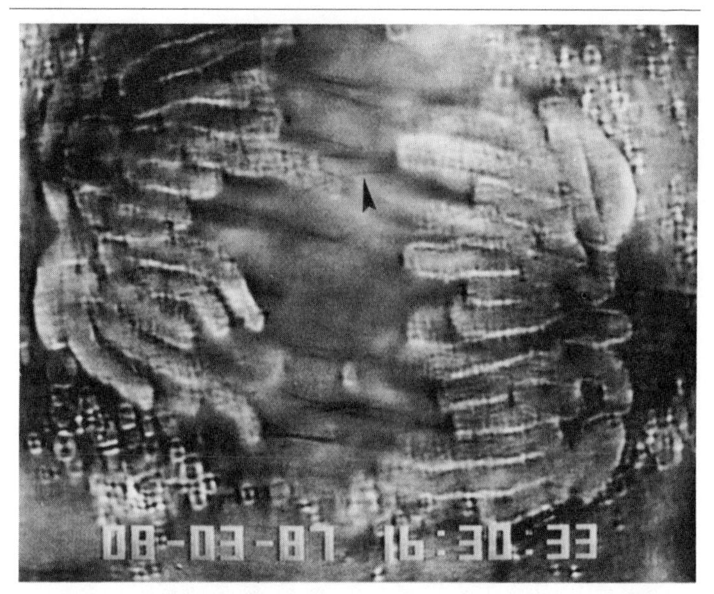

Figure 6. *Live newt lung epithelial cell in division observed with a high resolution polarizing microscope. The time-lapsed video record, from which these scenes were extracted, clearly shows the behavior of individual microtubules (arrow). Before digital enhancement and thin optical sectioning, the birefringence of single microtubules had not been seen in living cells. (From Cassimeris* et al., *1988.)*

that much, but it shows that our ability to visualize such thin objects has been dramatically improved by the use of video and digital image processing.

During my Centennial Evening Lecture, I projected, for the first time in public, some stereoscopic video, high resolution microscope images (Fig. 7). The technical details of the method is described elsewhere (Inoué, 1990a), but the intent, or aim, of our stereo images is to get a four-dimensional display, that is, three-dimensional images changing with time as the cell divides or as the embryo develops. It will still be some time before we can actually get the four-dimensional images, but the stereo video images that we have been able to generate so far should give you a sense of ongoing progress.

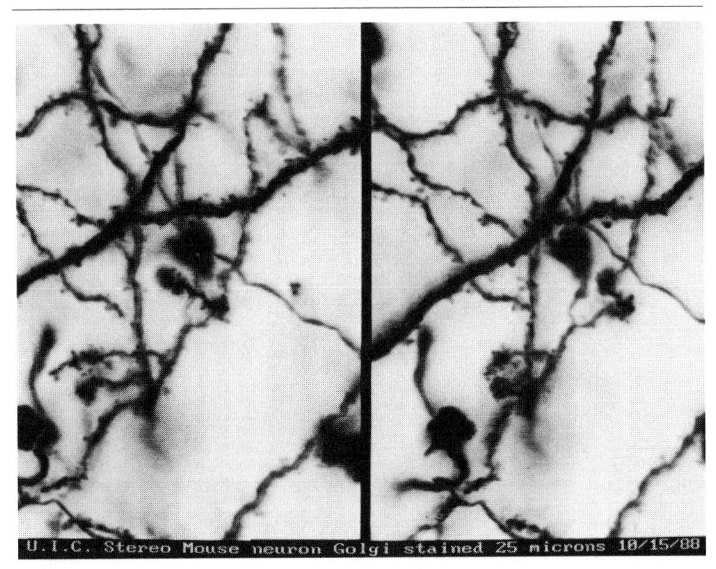

Figure 7. *Stereoscopic image of embryonic mouse neurons. Digitally enhanced optical sections of the silver-stained neurons were first stored sequentially in a laser disk recorder. They were then converted to stereo pairs with the Image-1/AT digital image processor. To see the stereoscopic image, view the pair with your eyes crossed. (From Inoué and Inoué, 1989. Microscope specimen courtesy of Steve Senft, Washington University, St. Louis).*

I'd like to conclude with the following summary and remarks. Unlike most man-made devices, the living cell is not patterned with a fixed architecture. To be sure there are guiding principles, building blocks, and a few permanent parts, but we are finding with increasing frequency that the building blocks are repeatedly taken apart and then re-assembled again into new structures to be used in different ways according to the functional needs of the cell.

Even the act of disassembling and reassembling the structures can play an important role in the cell's vital functions. For example, the assembly and disassembly of the nanometer-thick microfilaments and microtubules can change cell shape, give rise to tissue and cell polarity, move cells, transport organelles, direct traffic, and send signals within cells.

A precisely orchestrated building and taking apart of such dynamically shifting architecture is crucial for the cell, especially when it is producing its daughter offsprings. In cell division, growing and collapsing microtubules must precisely partition the gene-bearing chromosomes into the daughter cells. Repeatedly growing and shortening microtubules position the spindle, whose location and orientation in early anaphase signal the actomyosin where to assemble and cleave the cell (Inoué, 1990b). In this way the chromosomes are placed into the proper cytoplasm that govern the daughter cells' developmental fate.

To explore such intricate architecture of the cell, electron microscopists take advantage of the very high resolution at their disposal, chemists isolate and measure well-defined molecular reactions, and geneticists and immunologists make use of the exquisitely sensitive and specific biomolecular probes. Meanwhile, others use the light microscope to track dynamic changes in the whole living cell or molecular assemblies, as they actually take place.

Dr. Keith Porter, whom we honor, has done much pioneering work exploring the fine architecture of cells with the electron microscope. My students and I have had the pleasure of participating in the development, as well as use, of the light microscope to explore directly the dynamic behavior of the cell and its molecular parts.

With these complementary approaches, each with its unique strengths, albeit also with its limitations, we are gradually coming to comprehend the incredibly dynamic molecular architecture that underlies the miraculous organization, reproduction, and function of the miniscule living cell, on whose proper performance we, and all living creatures, each depend for physical and mental activities, for health, and for the propagation of life itself.

Several of the motion pictures and video scenes mentioned in this article are available from Alan Liss, Inc., as a 1/2 inch videotape supplement to the journal *Cell Motility and Cytoskeleton.*

LITERATURE CITED

Allen, R. D. 1985. New observations on cell architecture and dynamics by video-enhanced contrast optical microscopy. *Annu. Rev. Biophys. Biophysical Chem.* **14:** 265.

Allen, R. D., J. L. Travis, N. S. Allen, and H. Yilmaz. 1981a. Video-enhanced contrast polarization (AVEC-POL) microscopy: a new method applied to the detection of birefringence in the motile reticulopodial network of *Allogromia laticollaris. Cell Motil.* **1**: 275.

Allen, R. D., N. S. Allen, and J. L. Travis. 1981b. Video-enhanced contrast, differential interference contrast (AVEC-DIC) microscopy: a new method capable of analyzing microtubule-related motility in the reticulopodial network of *Allogromia laticollaris. Cell Motil.* **1**: 291.

Ashley, C. A., K. R. Porter, D. E. Philpott, and G. M. Hass. 1951. Observations by electron microscopy on contraction of skeletal myofibrils induced with adenosinetriphosphae. *J. Exp. Med.* **94**: 9-20.

Cassimeris, L., S. Inoué, and E. D. Salmon. 1988. Microtubule dynamics in the chromosomal spindle fiber: analysis by fluorescence and high-resolution polarization microscopy. *Cell Motil. Cytoskel.* **10**: 185-196.

Franzini-Armstrong, C., and K. R. Porter. 1964. Sarcolemmal invaginations constituting the T system in fish muscle fibers. *J. Cell Biol.* **22**: 675-696.

Huxley, A.F., and R. Niedegerke. 1954. Interference microscopy of living muscle fibres. *Nature* **173**: 971-973.

Huxley, A.F., and R.E. Taylor. 1958. Local activation of striated muscle fibers. *J. Physiol. London* **144**: 526.

Huxley, H., and J. Hanson. 1954. Changes in the cross-striations of muscle during contraction and stretch and their structural interpretation. *Nature* **173**: 973-976.

Inoué, S. 1952a. The effect of colchicine on the microscopic and submicroscopic structure of the mitotic spindle. *Exp. Cell Res. Suppl.* **2**: 305-318.

Inoué, S. 1952b. Effect of temperature on the birefringence of the mitotic spindle. *Biol. Bull.* **103**: 316.

Inoué, S. 1953. Polarization optical studies of the mitotic spindle. I. The demonstration of spindle fibers in living cells. *Chromosoma* **5**: 487-500.

Inoué, S. 1964. Organization and function of the mitotic spindle. Pp. 549-598 in *Primitive Motile Systems in Cell Biology*, Allen and Kamiya, eds. Academic Press, New York.

Inoué, S. 1981. Video image processing greatly enhances contrast, quality, and speed in polarization based microscopy. *J. Cell Biol.* **89**: 346-356.

Inoué, S. 1986. *Video Microscopy*. Plenum Press, New York.

Inoué, S.1987. Video microscopy of living cells and dynamic molecular assemblies. *Applied Optics* **26**(16): 3219-3225.

Inoué, S. 1988. The living spindle. *Zool. Sci.* **5**: 529-538.

Inoué, S. 1990a. Whither video microscopy? Towards 4-D imaging at the highest resolution of the light microscope. Pp. 497-511 in *Optical Microscopy for Biology*, Herman and Jacobson, eds.Wiley-Liss, New York.

Inoué, S. 1990b. Dynamics of mitosis and cleavage. Pp. 1-14 in *Cytokenesis:Mechanisms of Furrow Formation During Cell Division*, Conrad and Schroeder, eds. *Ann. N. Y. Acad. Sci.*, Vol. 582.

Inoué, S., and T. Inoué. 1989. Video enhancement and image processing in light microscopy. Part 2: Digital image processing. *Am. Laboratory* **April 1989**: 62-70.

Mazia, D., and K. Dan. 1952. The isolation and biochemical characterization of the mitotic apparatus of dividing cells. *Proc. Natl. Acad. Sci.* **38**(9): 826-838.

Mittal, B., J. M. Sanger, and J. W. Sanger. 1987. Visualization of myosin in living cells. *J. Cell Biol.* **105**: 1753-1760.

Porter, K. R. 1942. Developmental variations resulting from various associations of frog cytoplasms and nuclei. *Trans. N.Y. Acad. Sci.* **4**:213-217.

Porter, K. R. 1966. Cytoplasmic microtubules and their functions. Pp. 308-354 in *Principles of Biomolecular Organization*, G.E.W. Wolstenholme and M. O'Connor, eds. J. and A. Churchill Ltd., London.

Porter, K. R., and F. L. Kallman. 1952. Significance of cell particulates as seen by electron microscopy. *Ann. N.Y. Acad. Sci.* **54**: 882-891.

Sabatini, D. D., K. Bensch, and R. J. Barrnett. 1963. Cytochemistry and electron microscopy. The preservation of cellular ultrastructure and enzymatic activity by aldehyde fixation. *J. Cell Biol.* **17**:19-58.

Schrader, F. 1953. *Mitosis*, 2nd edition. Columbia University Press, New York.

Tilney, L. G., and S. Inoué. 1982. The acrosomal reaction of *Thyone* sperm. II. The kinetics and possible mechanism of acrosomal process elongation. *J. Cell Biol.* **93**: 820-837.

Tilney, L. G., and K. R. Porter. 1965. Studies on microtubules in *Heliozoa* I. *Protoplasma* **60**: 21-344.

Winthrop John Vanleuven Osterhout
(1871 - 1964)

Channels, Pumps, and Osmotic Machines in Plants: A Tribute to Osterhout

Introduction by Lorin J. Mullins

Although trained as a classical botanist, **W. J. V. Osterhout** concluded that what he was really interested in was how cells worked. To pursue such studies, he found, at a rather late date, that he had to learn physics and physical chemistry. And he did.

This change in his research interest can be seen by comparing his first paper, published in 1896, titled "Life history of *Rhabdonia*," with those published in 1906, in which he writes on the role of osmotic pressure in marine plants. Osterhout was one of the first to appreciate the value of electrical measurements in plants, and, as witnessed by the research interests of Dr. Clifford Slayman, this remains an important subject. While the studies of Osterhout were extensive, he was primarily interested in how an electrical potential difference could develop across the surface membrane of cells (animal physiologists were also most interested in this point, but there was no generally accepted theory in the 1930s).

An electrical potential difference across the cell membrane can be thought of as a kind of a suction that selectivity attracts positively charged ions at the outside of the cell to the inside. Why is this electrical potential difference important for plant cells? Plant cells live in what amounts to fresh water, yet the principal intracellular cation is potassium, which is present outside the cell at very low levels. Osterhout was convinced that the membrane potential was caused by the unequal distribution of potassium between the inside and the outside of the cell, a view that is accepted today for many cell types.

Osterhout was particularly interested in a freshwater plant cell called *Nitella* because it has an action potential that lasts for many seconds. This electrical disturbance propagates at about 1 cm/s, or less than 1/100th that of animal nerve fibers. Many of Osterhout's studies, from the late 1920s to the late 1930s involved measuring the electrical properties of this cell.

Osterhout had many collaborators at the Rockefeller Institute in New York and, at Harvard, many graduate students. Two of his most distinguished Ph.D. students were Wallace Fenn, who for many years was Professor and Chairman of Physiology at Rochester and Lawrence Blinks of Stanford. Both men became members of the National Academy of Sciences, and both carried out highly distinguished research programs in animal and plant physiology.

I suspect that I was asked by the Organizing Committee to make this introduction of Osterhout because I am one of the few individuals still around who actually met him in the late 1940s and early '50s. I remember well visiting him in his office at Rockefeller. He was retired, but this merely gave him the time to produce a great outpouring of papers.

I have tried to compose a summary about Osterhout but I find that the best one is by L. R. Blinks, in his biographical memoir of Osterhout published by The National Academy of Sciences in 1974. I am most grateful for the help of Professor Blinks in assembling this presentation. His summary statement is as follows:

"Winthrop John Vanleuven Osterhout was born in Brooklyn on August 2, 1871, a little over a century ago. He died in New York April 9, 1964. Elected to the National Academy of Sciences in 1919, he lived to be one of the older members (aged ninety-two). He greatly influenced the course of biology in the United States, as it turned from a largely descriptive into an experimental and analytical science. He was one of the founders of the new discipline of general physiology through his own work and through his editorship of the *Journal of General Physiology*, which he founded with Jacques Loeb in 1918. He remained an editor for over forty-five years and trained many students who contributed to general physiology."

Clifford Slayman received his undergraduate degree from Kenyon College in 1958 and his Ph.D. in physiology from Rockefeller University in 1963. At present, Dr. Slayman is a professor in the Department of Cellular and Molecular Physiology at Yale University. He is director of graduate studies in the Department of Physiology at Yale University, an editorial board member of the *Journal of Membrane Biology*, and a member of a number of Foundation advisory panels. He is also a member of the Biophysical Society, the Society of General Physiologists, and the American Physiological Society. I cannot claim that Clifford Slayman needs no introduction, because that would imply that I am superfluous, but his field is both sufficiently important and unknown that I will venture some introduction as to why the work is important.

We return to the membrane potential that so interested Osterhout. In the 1950s it became possible to give a coherent picture of how this electrical asymmetry arose at least in animal cells. The answer was that the potential corresponded to how much potassium was inside the cell as compared with the amount outside. This was how the idea that potassium ions were at equilibrium across the cell membrane originated. The potential was not quite this value because of a small leak of other ions that could be corrected for. This was dogma for a few years, but then it was discovered that metabolism—the process that keeps things running in cells—could produce a small but measurable contribution to the potential. Naturally, many individuals wanted to know whether metabolism might come to dominate this all-important membrane potential. Two papers here seemed important to me. In 1968 Kitasato showed that more than half of the membrane potential of *Nitella* (Osterhout's favorite algae) was controlled by metabolism rather than by the potassium ion gradient of the cell. More importantly, in 1965, Slayman was able to show that the very high membrane potential of *Neurospora* (the bread mold) that, at a maximum, could have a value of 200 mV could be virtually abolished by measures that reduced metabolism to low values. What did these findings mean? Nearly everyone in animal physiology was happy with the idea that potassium ions were virtually at equilibrium across the cell membrane, and yet here was a group of scientists saying that this is simply not so.

To be fair, I should say that the biochemists had been saying for many years that the hydrogen ion fluxes generated very large membrane potentials but the physiologists had replied that this only occurs in funny structures inside cells like mitochondria. How this all comes out is the subject of this chapter and I shall let Clifford Slayman tell you what the solution is.

Channels, Pumps, and Osmotic Machines in Plants: A Tribute to Osterhout

Clifford L. Slayman

Yale University School of Medicine

THE SCIENTIFIC CAREER OF W. J. V. OSTERHOUT spanned almost 70 years, from his undergraduate work at Brown University (1889-1893) to his dedicatory chapters in *Annual Reviews of Plant Physiology* (1957) and *Annual Reviews of Physiology* (1958). Although best known for his experimental studies in plant physiology and in membrane permeability phenomena, he made two other major contributions of a more general nature to biology. The first was to co-found, with his friend and mentor Jacques Loeb, the *Journal of General Physiology*, of which he remained an editor for more than 45 years. The second was to spur application of the physical and chemical sciences to biological problems. He did this both by the example of his own research: pioneering, *e.g.*, in the concepts of steady-state process and molecular carriers, and by enticing physical chemists to work in biology: specifically, he brought together the very active group (including MacInnes, Shedlovsky, and Longsworth) at the Rockefeller Institute. Although Osterhout was beset by health problems—including encroaching blindness—from the early 1930s onward, his knowledge and intellectual vitality continued to invigorate research for more than two additional decades.

Toward the end of that period his second wife and scientific collaborator, Marion Irwin Osterhout, befriended several students in the then fledgling graduate program of the Rockefeller University, whom she called "Dr. Osterhout's last students." The group included this author, along with Carolyn Walch Slayman, Alan Finkelstein, Frederick Dodge, and Roger Thies. Following her husband's lead, Marion Irwin too was a garrulous, entertaining, and broadly interested scientist, who profoundly influenced students. It was therefore a special pleasure to prepare this essay in honor of Osterhout, who was—at the beginning of my own scientific career—a personal living legend: of science, of the Rockefeller Institute, and of the MBL. The evening of this lecture itself also became a special event, because of the unexpected presence of six members of W.J.V. Osterhout's family: his younger daughter, Olga Osterhout Sears; his twin grandchildren, Harold H. Sears and Olga S. Karplus; great grandchildren Wendy S. Child, Kirsten Karplus, and Eric Karplus; and several husbands and wives. Also present were Mrs. S.E. Hill, and Dr. Robert Hill, the widow and son of S.E. Hill, who worked with Osterhout between 1929 and 1940; and Mr. John Valois, a well-known contemporary figure at MBL who collected *Nitella* for Osterhout. Their warm and enthusiastic presence very touchingly brought together two different and long separated aspects of the life of this very great man.

Historians with a penchant for assigning "age" names to particular periods of human existence will in the future very likely refer to the present time as the Age of Molecular Biology. It is an age in which the immensely powerful techniques of gene technology and physical biochemistry will yield great practical benefits—medical, agricultural, ecological—and will also reveal a detailed and fundamental unity underlying the vast diversity of phenomena that biologists have described and compiled over the past three centuries. But we are now just entering the age of molecular biology, and can look both backward and forward with special clarity. Thus, it is timely, and especially appropriate in honor of W. J. V. Osterhout, to summarize certain major features of functional organization in plant cell membranes, and to describe how those features relate to several important aspects of plant behavior.

Introduction

The *obvious* difference between animals and plants (and fungi and bacteria)—that plant cells are surrounded by a permeable but constraining wall of cellulose and other polymers—also happens to be a *fundamental* difference. Among their many attributes, all living cells are solution droplets of macromolecules (proteins and nucleic acids) surrounded by a "slightly" permeable surface membrane, which allows small ions and nutrient molecules, but not the essential macromolecules, to pass. These large indiffusable molecules have both an osmotic effect and an electrical effect, which would—if unchecked—lead to continuous entry of salt and water to the cell interior, thus diluting the macromolecules and stretching the surface membrane to the breaking point.

Cell walls are one of evolution's two major devices to combat physical dissolution: walls simply allow sufficient hydrostatic pressure (turgor) to develop within plant cells to block net entry of water. The second major device is the use of metabolic energy specifically to export excessive ions (such as sodium), which thereby become an *extra*cellular osmotic balance to the macromolecules and ions inside. Evolutionary divergence following upon these two distinct options has been spectacular. The first device, which greatly restricts cell-to-cell communication and long-distance locomotion, supported development of the massive structures and enormous surfaces required for large-scale solar energy conversion: that is, our grasslands and both marine and terrestrial forests. But the second device, which cannot support massive structures except in the marine environment, did allow both intimate cell-to-cell communication and long-distance locomotion, eventually leading to intelligence and to complex sentient organisms.

It was one of Osterhout's great articles of faith, that proper application of the quantitative tools of physical chemistry to the vast divergence of problems in biology (Osterhout, 1922), including problems of osmotic balance in plants and animals, would reveal an underlying mechanistic unity. His most important technical contribution grew out of that faith: the systematic use of electrical techniques to assess permeability, an advance spawned by the rapid progress in electrochemistry at the end of the 19th century (Nernst, 1890; Planck, 1890). Osterhout realized that measurements of voltages, resistances, and solute compositions *across* cell surfaces were essential for understanding ionic processes in living cells, and also that the giant cells of characean algae would provide excellent material for such measurements (Osterhout, 1927). The fresh water alga *Nitella*, first collected by Osterhout from Nobska pond (and still collected for him by Mr. Valois into the 1950s) became his favorite material. The great length (10 cm or more) of its internodal cells permitted simple estimation of electric potentials across the surface by means of exterior, longitudinal recording of demarcation potentials or currents. This technique, which had been used for studies of conduction in muscle fibers and nerve axons (Steinbach, 1896; Hermann, 1905; Alcock, 1906), is shown diagramatically in Figure 1, along with several more recent techniques. About the same time, Osterhout

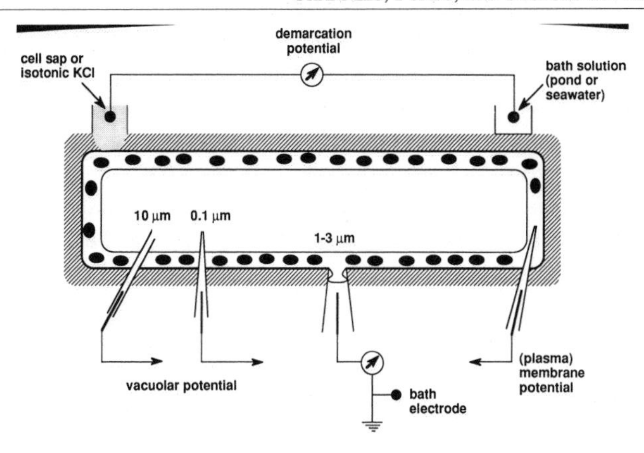

Figure 1. *Diagram of a characean algal cell (from Peebles* et al., *1964), showing different kinds of electrodes that have been used to measure the electric potential difference across the cell surface. The overall length of such a cell can be 10 cm or more; and diameter, 1 mm or more. More than 90% of cell volume is occupied by a central vacuole, containing fluid "cell sap," which is usually acidic and contains enough salt (organic ions, plus Na^+, K^+, and Cl^-) to be in osmotic equilibrium with cytoplasm. The vacuole is bounded by the tonoplast (——), a well-defined unitary membrane, and is, in turn, surrounded by cytoplasm containing the usual plant-cell organelles, of which the most prominent are the chloroplasts. The cytoplasmic surface is formed by another unitary membrane, the plasmalemma (——), surrounded by a cellulose wall (cross-hatching) capable of withstanding turgors of 10-15 atmospheres. Membrane thicknesses are about 50 Å; average cytoplasmic thickness, 10-15 μ; and cell wall thickness, several μ.*

Recording arrangements—(Top) Plasmalemma is locally "short circuited" by an electrode (left-hand) containing cell sap, so the inside-to-outside difference of electric potential can be estimated at a remote electrode (right-hand) holding pondwater (Osterhout and Harris, 1927). (Bottom) Newer recording modes (read left to right): insertion of a large capillary pipette directly into the vacuole (Taylor and Whitaker, 1925; Osterhout, 1927); insertion of a microcapillary into the vacuole (Etherton and Higinbotham, 1960); tight attachment of a capillary electrode to cleaned membrane (Neher and Sakmann, 1976; Hamill et al., *1981); insertion of a microcapillary into cytoplasm (Walker, 1955; Slayman and Slayman, 1962; Bates* et al., *1982).*

(1927) and Taylor and Wittaker (1925) simultaneously devised capillary electrodes that could be inserted into the vacuoles of giant algal cells. Osterhout and his students applied this technique to several giant-celled algae, including especially the marine species *Valonia* (reviewed in Osterhout, 1931).

True microcapillaries were developed 20 years later (Graham and Gerard, 1946), again for use on nerve and muscle, but were soon applied widely to plant cells (Walker, 1955; Etherton and Higinbotham, 1960; Slayman and Slayman, 1962). However, because the cytoplasmic layer in mature plant cells is rather thin and because the wall of a turgid cell acts like a drum head, entry of pipettes occurs suddenly and usually punches through both membranes and into the vacuole. With special care (Walker, 1955; Bates *et al.*, 1982), it has been possible to insert a pipette just through the plasma membrane, and the stable voltage thus measured is now commonly termed the cell's *membrane potential.* Finally, the past few years have seen a burgeoning use of so-called patch electrodes (Neher and Sakmann, 1976; Hamill *et al.*, 1981): polished capillaries *ca.* 1 mm in diameter, which—under gentle suction—attach so tightly to naked cell membranes that currents as small as ~0.1 pA (10^{-13} ampere) can be measured flowing through individual channel-molecules in the membranes. This technique looks especially powerful, and is now generating important light, along with spectacular heat, in plant biophysics (Raschke *et al.*, 1988; Schroeder and Hagiwara, 1989; Hedrick and Schroeder, 1989).

The Origin of Bioelectric Potentials

Between 1925 and 1945, Osterhout and his students published well over 100 research papers that described investigations of ionic selectivity at the *Nitella* surface. In most cases they monitored the influence of bath composition upon demarcation potentials. Variations of the major physiological ions were tested: sodium, potassium, chloride, and calcium; lipid solvents, vital dyes, and metabolic inhibitors were also explored. The understanding reached from this work (Osterhout and Hill, 1936), in consensus with several laboratories studying nerve and muscle tissue (Amberson, 1936; Cole, 1940; Goldman, 1943), was that bioelectric potentials arise from differential diffusion of ions between the interior and exterior of cells, depending both upon gradients maintained by metabolism and upon selective permeability of the cell surface.

Table I

Ionic parameters (approximate) for Nitella[a]

Ion	Concentration (m*M*)	Diffusion[b] potential (mV)	Measured[c] potential (mV)
CYTOPLASM			
Potassium	**125**	**-180 to -200**	
Sodium	**10**	**-30 to -90**	
Chloride	**15**	**+30 to +155**	
VACUOLE			**-140 to -180**
Potassium	**80**	**-170 to -190**	**resting**
Sodium	**30**	**-58 to -115**	
Chloride	**125**	**+80 to +210**	**-10 to -40**
			action pot.
BATH			
Potassium	**.05 - .1**		
Sodium	**.3 - 3.**		
Chloride	**.03 - 5.**		

[a]Roughly averaged values summarized mainly from Hope and Walker (1975).
[b,c]Voltages calculated (b) and measured (c) as vacuole minus bath.

Following the original suggestion of Bernstein in 1902, cell surfaces were viewed as potassium-specific electrodes, through which that ion could diffuse rapidly outwards, whereas other ions were retarded. The greater escaping tendency of K^+ would create an electric potential difference—cell exterior positive—which ultimately must limit the net escape of K^+ and couple it to the slower movement of anions. For giant algal cells, this view accorded with the normal distributions of ions between cell sap and the exterior (Table I), with available estimates of the potential difference, and with the manner in which potential varied as the ionic composition of the bathing solution was changed. The concept seemed proven after Hodgkin and Huxley (1952) presented their quantitative description of the nerve action potential in terms of a large, transient increase in membrane permeability to sodium, which—being more concentrated outside cells than inside—would temporarily drive the axon interior electropositive.

Action potentials had also been discovered in plant cells (see, *e.g.*, Burdon-Sanderson, 1883; Snow, 1924), and Osterhout and his students described those of *Nitella* in great detail (Osterhout, 1934, 1946). (A train of spontaneous action potentials in *Nitella*, observed by Osterhout via the demarcation technique, is shown in Figure 5A.) They are "proper" action potentials, displaying a threshold, all-or-nothing response, spontaneous repolarization, and a refractory period, and representing a decrease in both membrane potential and membrane resistance (Cole and Curtis, 1938). In several important respects plant action potentials differ conspicuously from nerve action potentials: they are somewhat more variable in waveform, much longer (1-100 s) in duration than conventional nerve action potentials (1-10 ms), and reflect a large transient increase of membrane permeability to *chloride* (Gaffey and Mullins, 1958; Mailman and Mullins, 1965), rather than to sodium.

A summary of the major ion diffusion processes that determine bioelectric potentials in plant and animal cells is shown in the diagrams of Figure 2. Major efforts over the past 30 years have been directed toward documenting passive ion permeabilities, and fluctuations thereof, in increasing numbers of plant species. Of course the analogy with porous-barrier diffusion has been replaced by the notion that individual ions pass through molecular channels that are rather specific for each ion species; and an increasing catalogue of ion channels is coming to be recorded as individual molecules by means of the patch electrode technique. More than two dozen different, discrete channels have been isolated and studied in patches of plant cells membranes (see, *e.g.*, Hamill *et al.*, 1981; Schroeder *et al.*, 1984). Most thus far have proven to be potassium selective (Laver and Walker, 1987; Blatt, 1988; Schroeder, 1988), and a concensus is growing that these K^+ channels are the principal agents of resting membrane permeability. A clear demonstration of this arrangement has been presented for the characteristic K^+ channels in a giant marine alga, *Acetabularia* (Bertl and Gradmann, 1987).

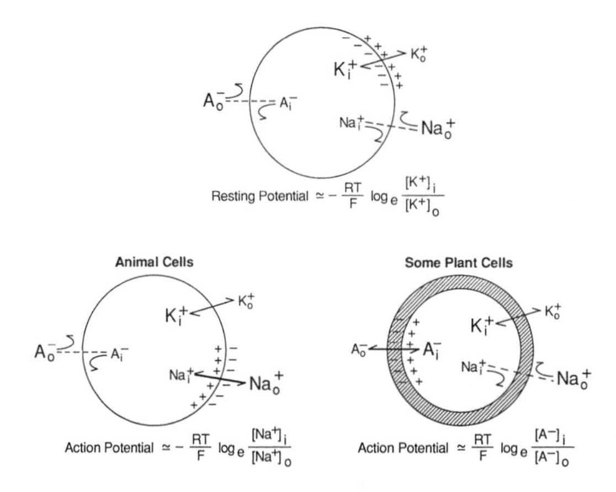

Figure 2. *Major ion-diffusion regimes in living cells, and their relationship to membrane potentials. (Top) For most cells the unperturbed membrane has much higher permeability to potassium than to other ions, and its tendency to escape from the high cytoplasmic concentration leaves cell interiors electronegative. (Bottom) During action potentials, most animal cells become transiently permeable to sodium, which tends to flood inward, driving the cell interior electropositive. In plant cells, permeant anions tend to escape from the cytoplasm/vacuole, which also leaves the cell interior transiently less negative than normal.*

A sample recording from one such channel is displayed in Figure 3A. The patch pipette was pressed against naked membrane reformed on a protoplasmic droplet that had been released into seawater by slashing the cap of a mature plant. When slight suction was applied to the membrane, it sealed to the pipette with a resistance of ~10 GΩ (10^{10} ohms). Thereafter, the voltage at the electrode tip was electronically clamped in steps of 10 mV, for intervals of ~30 s, and the current required to hold each voltage was recorded. Single channel openings appear as sudden deflections of the trace, during which several 100 million ions/s can rush through. Only very small currents were observed near -30 mV, which represents diffusion equilibrium for potassium. But currents became large and positive (upward) for voltages positive to -30 mV, or large and negative for voltages negative to -30 mV. Also the active channel in this patch spent more time open as the potential was shifted positive.

Calculation of the average current for each recording interval yielded the plot and fitted curve in Figure 3B, which closely resembles the behavior of the whole plasma membrane in *Acetabularia* (Gradmann, 1975). The steep rise of current at positive voltages means that potassium exit from the cells is disproportionately easy during strong depolarization. Animal cells usually avoid this feature of potassium channels, but it is very useful and important for many plant cells.

Active Transport Processes in Plants

Of course much of Osterhout's attention—and that of many other physiologists, as well—was devoted to the obvious need for coupling uphill ion movements to metabolic reactions to create the steady gradients of potassium, sodium, and chloride, which support the downhill (diffusion) mechanisms. The postulated coupling process became known as *active transport* or *ion pumping*, but its nature remained conjectural until the late 1950s, when two important discoveries were made: (i) ion pumping was identified in the major energy-producing organelles: mitochondria and chloroplasts (Bartley and Davies, 1954; Brierley, 1976), and (ii) active transport of sodium outward across animal plasma membranes was clearly identified with a distinct, isolatable, membrane-bound enzyme—the so-called Na^+, K^+-ATPase—which split ATP but required both cations as cofactors (Skou, 1957).

About the same time, general availability of fine microelectrodes elicited electrical studies on many kind of plant cells, mostly much smaller than *Nitella* or *Acetabularia*, and work in several laboratories (Higinbotham *et al.*, 1964; Slayman, 1965; Spanswick, 1967) revealed some surprising phenomena. For example, resting membrane potentials were often observed well beyond the maximal value for potassium diffusion (*ca.* -200 mV); and they were quite insensitive to extracellular concentrations of potassium, sodium, or other ions, but were quickly abolished by *metabolic* inhibitors. A clear demonstra-

---

Figure 3. *Elementary electrical properties of channels in the surface membrane of a protoplasmic droplet from the giant-celled marine alga,* Acetabularia. *(A) Record of a single channel obtained with a patch electrode, arranged in the "cell-attached"* mode (i.e., *as shown diagramatically in Fig. 1). Seal resistance: $8x10^9$ ohms; bathing solution and electrode solution: 130* mM *KCl. Patch voltages listed are corrected for the resting membrane potential of the droplet. (B) Plotted points represent the (clamping) current*

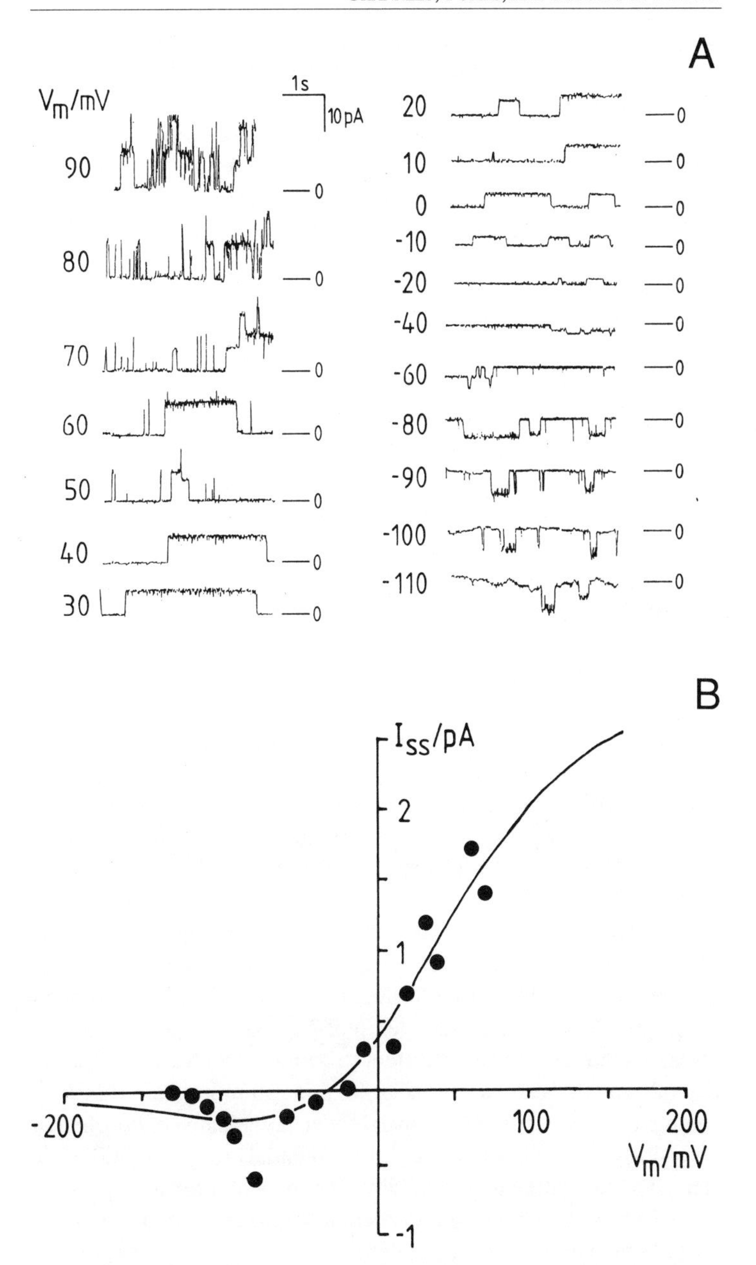

through the channel, averaged over each 30-s recording interval. The fitted current-voltage curve drawn closely resembles total potassium current through the intact plasma membrane of Acetabularia *(Gradmann, 1975). (Reproduced with permission from* J. Membrane Biol., *Vol. 99, pages 41-49. A. Bertl and D. Gradmann. 1987.)*

tion of this metabolic dependence was provided in a rapid depolarization of the *Neurospora* plasma membrane caused by brief pulses of sodium azide, as shown in Figure 4A (Slayman *et al.*, 1970).

That depolarization, from a resting potential near -200 mV to about -30 mV, was not accompanied by either an appreciable resistance change or an appreciable loss of cytoplasmic ions—as would have been expected for a shift in diffusion potentials, but was accompanied by rapid decay of ATP. It seemed, therefore, that metabolism itself could power an ion current through the membranes under steady-state conditions; or, in other words, that the enzymatic process occurring within ion pumps could displace separated charges, not just ion pairs, across membranes.

Thus, ion pumps were said to be *electrogenic*, a notion that had multiple historic origins and is very general in its implications. It is now known that almost all biological membranes, cellular and organellar, contain electrogenic ion pumps, and thereby function as fuel cells, able to convert the energy of chemical bonds directly into electrical energy, in a cyclic and fully renewable fashion. Identification of the fuels and current-carrying ions for such pumps has subsequently become a major research topic. For the plasma membranes of all eukaryotic cells, the fuel is ATP—the energy currency ubiquitous throughout biochemistry. In animal cell membranes, the ion current is sodium; but in the plasma membranes of most plants, bacteria, and fungi, the ion current is protons, although in some marine algae it is chloride (Goldfarb *et al.*, 1984).

Ion currents passing through individual pump molecules (10-1000 ions/s) are far too small to be measured yet by electrophysiological techniques, but the ensemble behavior of 10^5 pumps (or more) can be recorded, and can often be distinguished from membrane channels and other charge carriers by their rapid response to metabolic inhibitors. Figure 4B shows, in the "difference" curve, how current through the plasma-membrane proton pump of *Neurospora* varies with imposed potential. This result differs strikingly from that, *e.g.*, for the K^+ channels in *Acetabularia* (Fig. 3B), in saturating near 0 mV, in completely lacking the zero- or slightly negative-slope region, and in reversing only at very large negative potentials (beyond -360 mV), characteristic of the energy available from ATP hydrolysis rather than of the energy present in the transmembrane concentration difference for protons (Gradmann *et al.*, 1978).

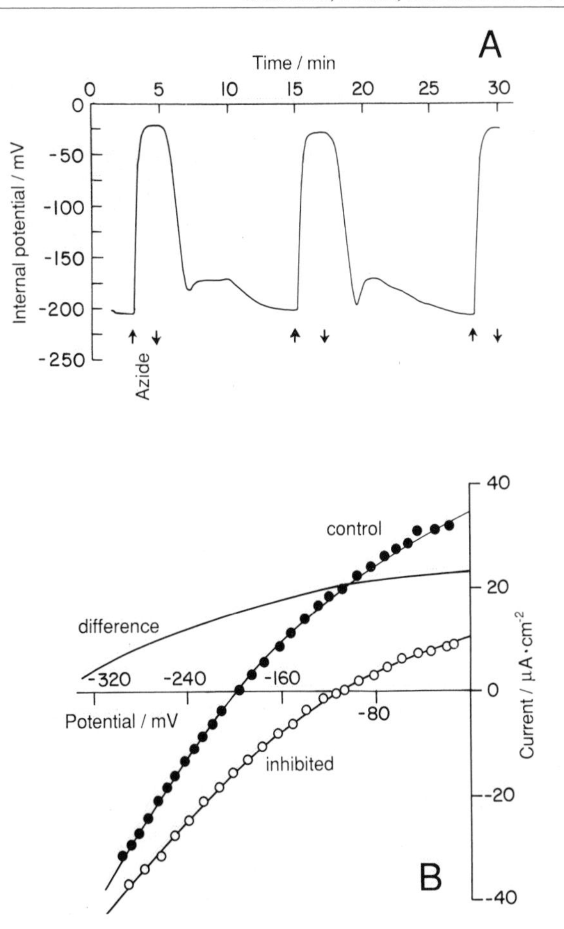

Figure 4. *Elementary electrical properties of the plasma-membrane proton pump (H^+-ATPase) in* Neurospora. *(A) Depolarization of the membrane during pulses of sodium azide (1* mM*). Up arrows: entry of inhibitor solution into the chamber. Down arrows: start of inhibitor washout. Note potential shift from* ca. *-200 mV to* ca. *-30 mV in the presence of azide (Slayman* et al.*, 1970). (B) Steady-state current-voltage relationships for the* Neurospora *plasma membrane in the absence (control) and presence (inhibited) of 1* mM *potassium cyanide. The difference (c-i) curve approximates the I-V curve of the proton pump (Gradmann* et al.*, 1978). Respiratory inhibition was less complete with cyanide than with azide.*

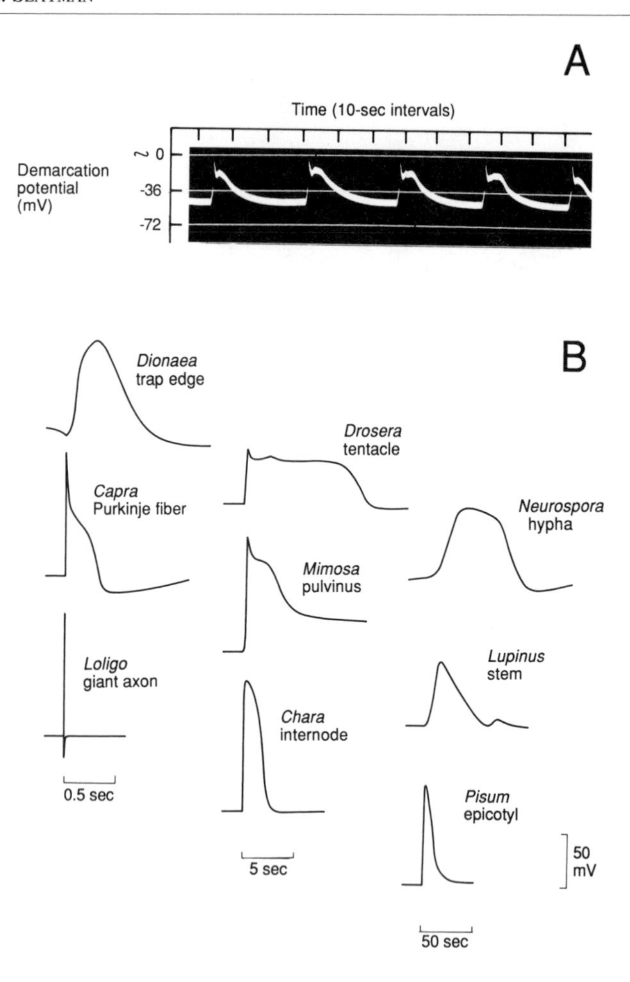

Plant Action Potentials

Although the mechanisms underlying ionic balance and surface selectivity occupied Osterhout's *experimental* attention, questions of biological purpose motivated his interest, as is attested by the recurrent theme of "The Work" (or function) of individual plant organs, in his highly successful textbook, *Experiments with Plants* (Osterhout, 1905). In the present context, a convenient way to introduce the roles of ion channels and pumps in plant *behavior* is to examine the possible functions of plant action potentials. Episodes of spontaneous depolarization, resembling nerve action potentials, were first observed (Burdon-Sanderson, 1873) in the trap organ of a carnivorous plant, the

Figure 5. *Representative plant action potentials. (A) A classic Osterhout record of* Nitella *action potentials (Osterhout, 1946). (B) Modern recordings of various plant action potentials, compared with the familiar events in cardiac Purkinje fibers and in squid giant axon. Uniform amplitude scale (lower right), but note shifting time scale at the bottom of each column.*

Species	**Common name**	**Recording method**	**Ionic conditions**	**Reference**
Dionaea	Venus flytrap	Leaf surface potential	Plant in air	Sibaoka, 1966
Capra	Goat	Penetrating electrode	Mammalian Ringer	Weidmann, 1957
Loligo	Squid	Penetrating electrode	Seawater	Moore & Cole, 1960
Drosera	Sundew	Surface of tentacle	Plant in air	Williams & Pickard, 1972
Mimosa	——	Penetrating electrode	Peeled pulvinus	Abe & Oda, 1976
Chara	——	Penetrating electrode	Pondwater	Findlay & Hope, 1964
Neurospora	Pink br. mold	Penetrating electrode	25 m*M* K^+, 1 m*M* Ca^{++}	Slayman *et al.*, 1976
Lupinus	Lupine	Double electr. on stem	Plant in air	Paszewski & Zawadzki, 1974
Pisum	Pea	Isolated epicotyl	10.1 *M* KCl	Pickard, 1971

Venus flytrap (*Dionaea*). And the obvious function of that action potential is the same as in animal cells: intercellular communication, and activation of the trapping mechanism.

A modern recording of the *Dionaea* action potential, via a wick electrode placed at the edge of the trap (Sibaoka, 1966), is shown in Figure 5B. Its amplitude, about 100 mV recorded extracellularly, is comparable with that of animal cells, but its duration—close to 1 s—is not, as can be seen from the accompanying drawings of a slow action potential in goat (*Capra*) Purkinje fibers, and the fast action potential in squid (*Loligo*) giant neurons. The *Dionaea* action potential is one of the briefest known in plants, even including other motor organs, such as the tentacle of sundew (*Drosera*) or the pulvinus of *Mimosa* (a mechanically sensitive plant). Action potentials in plants that lack obvious associated motions, such as *Nitella*, pea (*Pisum*), and *Neurospora*, are generally much longer, lasting from several seconds to well over one minute.

Because it is clear that plant action potentials are far more widespread than carnivorous behavior, it seems likely that their basic purpose is not communication, but their unavoidable physical consequence in plants: *viz.*, changes of volume or turgor. Action potentials usually represent an increased permeability, and in plants an increased tendency for net outward diffusion of ions, which can last a long time. The chloride action potentials of *Chara* or *Acetabularia*, for example, permit net efflux of chloride with potassium (see Fig. 3B; Mummert and Gradmann, 1976); accompanying water exit must decrease both cell volume and turgor pressure. Such expected reductions of turgor have in fact been measured for both species (Barry, 1970; Wendler *et al.*, 1983), in the latter case by means of a vacuolar micropipette adapted as a pressure transducer (Hüsken *et al.*, 1978). For *Acetabularia* maintained in a constant environment, turgor pressure rises and falls along a rhythmic sawtooth waveform, and each fall (about 0.3 atm) is accompanied by one or more action potentials. This and related observations have prompted the notion that action potentials occur in response to "excess" turgor or ion content, as a purposeful mechanism to unload ions (Mummert and Gradmann, 1976; Wendler *et al.*, 1983).

Plant cells, unlike animal cells, need such a regulatory mechanism, not just because of the cell wall, but because they often operate with very high membrane potentials: sufficient to drive permeant cations to physically absurd concentrations (*e.g.*, 10 *M*). High membrane potentials allow cells—especially free-living ones—to concentrate rarified nutrients, but also impose the need for (semi)periodic relief. This relief process itself has been used by evolution as the basis of action in osmotic machines.

The Osmotic Machines

The rigidity of plant cell walls seems to have prohibited evolution of actomyosin machines for extracellular movement, and plants developed osmotic machines instead. Two of the most interesting machines control transpiration and leaf position in higher plants, in relation to multiple environmental factors.

So-called *stomatal* pores—found most commonly on the undersides of leaves, and giving access to the interstitial space of the leaves—open and close via turgor/volume changes to regulate both water loss and leaf CO_2 tension for photosynthesis. The action of one type of pore, photographed from the leaf of the Asian day flower (*Commelina communis*), is demonstrated in Figure 6 (Willmer, 1983), with the pore shut (A) and then open (B). The two main regulatory cells, termed guard cells, are constrained—partly by their own wall structure and partly by the surrounding subsidiary cells—to behave rather like sausage balloons: if their ends are held at fixed spacing, as their internal pressure increases the balloons must pop apart in the middle; and then as pressure diminishes they can relax back to a slit. By means of specific ion-sensitive microelectrodes, vacuolar potassium concentrations have been determined for stomatal complexes in several different plants, including *Commelina*, with results as shown in Figure 6C and D (Penny and Bowling, 1974). For closed stomata, potassium concentrations are highest—the order of 450 m*M*—in the surrounding epidermal cells, and lowest toward the center (*i.e.*, in the

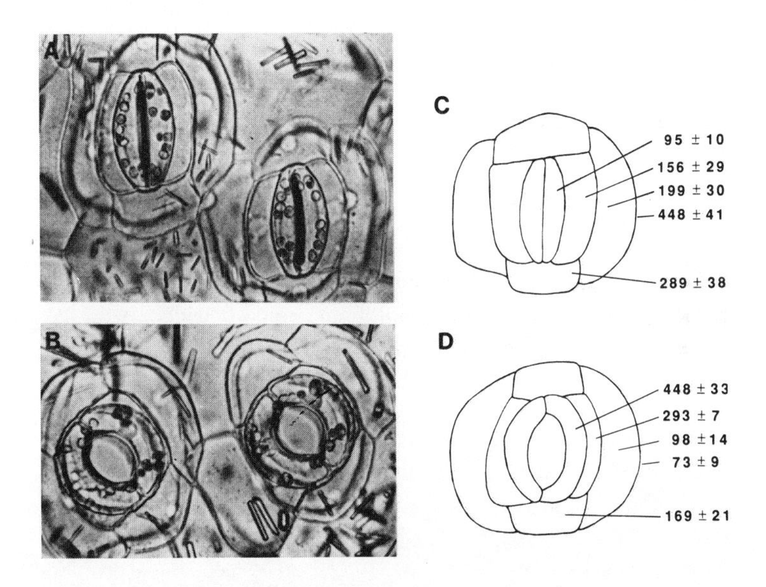

Figure 6. *Ion concentrations in relation to stomatal action. Photograph of epidermis stripped from leaves of the Asian day flower (*Commelina communis*), showing a pair of closed stomata (A) and a separate pair of open stomata (B), on the lower leaf surface. Photos courtesy of Dr. C.M. Willmer, University of Sirling, U.K. (1983). Potassium content of different cells in the stomatal complex, with the stoma closed (C) and open (D), redrawn from Penny and Bowling (1974)*[1].

inner subsidiary cells or the guard cells). For open stomata, the highest concentrations—again about 450 m*M*—occur in the guard cells, while much lower concentrations obtain in the surrounding epidermis. Although the whole balance sheet has not yet been drawn, this cation shift, along with accompanying anions, could easily account for measured pressure changes of about 10 atmospheres (Penny and Bowling, 1974) in open *versus* closed stomata[1].

A recent exciting development in stomatal biophysics has been the discovery, via patch electrodes, of potassium-specific channels in protoplasts of several different kinds of guard cells (see, *e.g.*, Schroeder *et al.*, 1984; Blatt, 1988; Schroeder, 1988). Some of these channels are voltage-gated and others are chemically modulated. So we can now sketch out specific models on how channel switching, against a background of electrogenic proton pumping, might bring about the observed potassium migrations. One of the most surprising facts to face, however, is that it is still not really known how cellular accumulation of potassium occurs in plant systems: whether by mechanisms linked to proton pumping via the membrane potential, by paired ion movement, or directly by ATP-linked enzymes. This and a number of other general problems in plant physiology may well be resolved via experiments on stomata.

Another important osmotic machine is that responsible for leaf closing and opening in night-closing (nyctinastic) plants, of which the best studied is *Samanea saman*, the individual leaflets of which fold against their pairs during the night-phase of the diurnal rhythm. The pulvinus, or motor organ for this folding, is a modified piece of the leaf stalk, shown in cross-section in Figure 7A, which consists of a vascular core surrounded by 10-20 layers of more-or-less spherical cells termed parenchyma (Satter *et al.*, 1974; Satter, 1979). The parenchymal cells are highly interconnected, particularly near the vascular bundles, and are organized into two regions, those that swell and become turgid as the leaves open—the extensor cells—and those that swell and become turgid when the leaves close—the flexor cells [illustrated in Fig. 7A (Satter *et al.*, 1982)]. (The mechanical problem of directed motion in this organ is different from that in stomata, in that it involves the entire macroscopic tissue.)

Freezing-point analysis of fluid droplets from different locations

[1]Some investigators have measured the lowest K^+ concentrations persistently in the inner lateral subsidiary cells for both open and closed stomata, but that does not vitiate the idea of ion flow between guard cells and their surrounds.

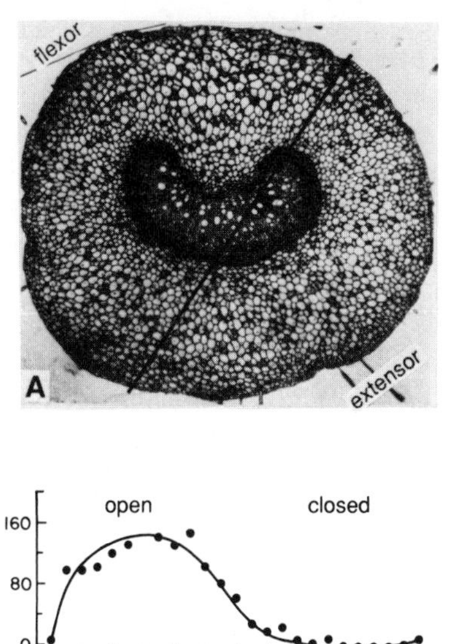

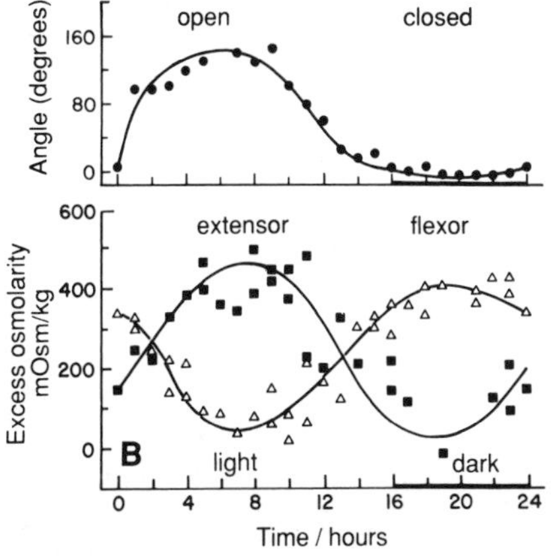

Figure 7. *Osmotic shifts in the pulvinus of a nyctinastic plant, associated with leaf movements. (A) Cross-section photograph of a primary pulvinus in* Samanea saman *(Satter, 1979). Central "bean" contains the vascular bundles; surrounding cells comprise the parenchyma. The diagonal line separates flexor (upper left) and extensor (lower right) components of the organ (Satter* et al., *1982). Reproduced with permission from Satter (1979). (B) Local osmotic content of the extensor and flexor regions, sampled from crushed tissue and assayed by freezing-point depression. Redrawn from Gorton (1987). Points plotted represent the difference between actual osmolarity at each time and the overall average tissue osmolarity. Note darkness from 16 h onward. (Reproduced with permission from* Encyclopedia of Plant Physiology, New Series Vol. 7: Plant Movements. *W. Haupt and M. E. Feinleib, eds. Springer-Verlag, Berlin. 1979.)*

in the pulvinus, as well as specific ion-mapping with an x-ray microprobe (Satter *et al.*, 1982; Gorton, 1987), have shown an excess of bulk ions (K^+ and Cl^-) located in the flexor region with leaves closed. During opening, 50-70% of K^+ and Cl^- migrates to the extensor region. This net shift of osmotic potential during the light-dark cycle is demonstrated in Figure 7B.

Many quantitative questions about ion migration within the pulvinus remain to be answered, both because the analytic techniques used thus far do not resolve ion contents of single cells and because the cell walls and interstitial space are an appreciable fraction of the total tissue volume[2]. Patch-recording experiments (Moran *et al.*, 1988) have now begun to identify K^+ channels in the membranes of protoplasts from pulvinus, but it is clear that much still remains to be learned, both about the types of channels present, and about phase relationships between channel- and pump-activities. Indeed, reciprocal control of ion loss and reaccumulation in the two opposing tissues of the motor organ is now becoming a major direction in research.

Initial Insights From Molecular Biology

The problems of surface structure, specificity, and transport—which were first addressed clearly in plants by W. J. V. Osterhout—are only just beginning to yield to the modern techniques of molecular biology. The major contribution molecular biology can make, at present, to such problems is determination of the primary amino acid sequences for intrinsic membrane proteins, from cloned cDNA for their structural genes; and the availability of primary sequences for families of functionally related proteins can lead to new hypotheses about evolutionary pathways, protein folding geometries, and the location of key functional groups.

Thus far, no genes for plant or fungal channel proteins have been reported cloned or sequenced. However, several genes for fungal and bacterial coupled-ion transporters have been sequenced, as have been more than two dozen genes for plant and fungal ion pumps. A feature common to all of the inferred membrane protein sequences—including those for channels, pumps, and coupled-ion porters in animal cells—is the presence of numerous and regular hydrophobic seg-

[2]For example, just how do intracellular and extracellular $K^+ + Cl^-$ relate to cell turgor and average tissue turgor? Or why are flexor-cell membrane potentials large when cells are K^+-depleted (Satter *et al.*, 1982)? High membrane potentials are usually a sign, in plants, of vigorous proton pumping, which should cause K^+ to be accumulated.

ments of amino acids that appear to form α-helices imbedded in the membranes. Channel molecules (all still from animal cells) seem to be multimeric, with compound clusters of the imbedded α-helices (Guy, 1984; Huganir, 1988). Some pumps, such as proton pumps in chloroplast, mitochondrial, bacterial, or vacuolar membranes, are also multimeric, but most ion pumps and coupled-ion porters appear to be much simpler: consisting of a single polypeptide of $M_r = 40,000$-150,000, which contains 6-12 transmembrane segments and the characteristic catalytic center (*i.e.*, nucleotide binding site, phosphorylation site, and hydrolytic cavity) for each enzyme.

The clearest picture has emerged for the proton pumps. cDNA sequences for five primary fungal enzymes and several plant enzymes have been published, along with sequences for several minor plasma-membrane ATPases. Inferred amino acid sequences for the *Neurospora* protein (Hager *et al.*, 1986) and those for two yeasts, *Saccharomyces cerevisiae* (Serrano *et al.*, 1986) and *Schizosaccharomyces pombe* (Ghislain *et al.*, 1987) are astonishingly alike: more than 70% identical among the three species, despite some wide evolutionary separation of the species in other respects. That finding suggests that at least one of the protein's functions, either proton extrusion *per se* or generation of the large membrane potential, has been very important in evolution.

A speculative cartoon diagram of the plasma-membrane H^+-ATPase of *Neurospora* is presented in Figure 8. The constituent features of this picture are based on freeze-fracture electron microscopy (Slayman *et al.*, 1989), physical studies (Hennessey and Scarborough, 1988), immunoblots (Mandala and Slayman, 1988), and calculations from the primary amino acid sequences. The detailed arrangement of the those features has not yet been determined. From hydropathy analysis (Engelmann *et al.*, 1986; Hager *et al.*, 1986), the enzyme contains four membrane-embedded segments, each consisting of two antiparallel α-helices connected by a short extracellular link (helices numbered **1,2,...,7,8** in Fig. 8). Together, these comprise about 25% of the total amino-acids. The remainder of the molecule is strongly hydrophilic and protrudes into the cytoplasm. It consists of the N-terminus (115 amino acids preceding Helix-1: $\mathbf{T_N}$ in Fig. 8), which has a large excess of negative amino acids (15-21) in the known fungal species; a loop (123 aa's from H2 to H3: **D** in Fig. 8) which has been suggested to bind the transported ions; a very long loop with several subdomains (361 aa's from H4 to H5: **P,N,H**), which bind nucleotides and contain the phosphorylated residue (Asp378); a short

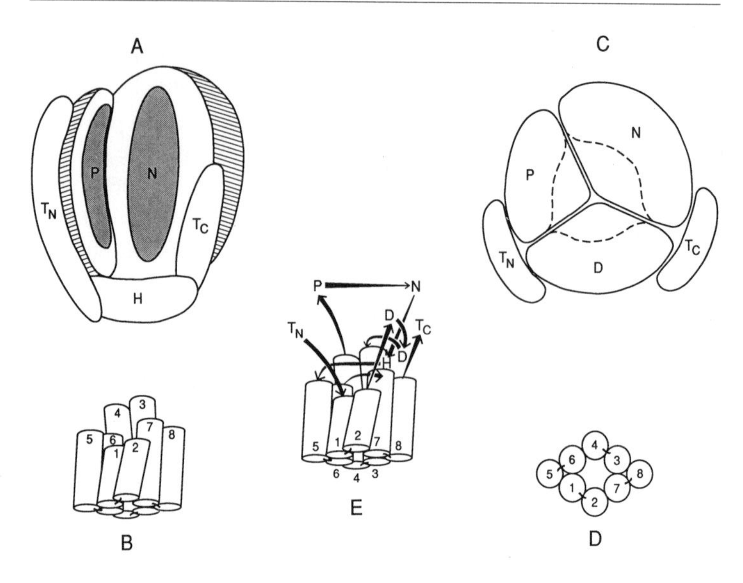

Figure 8. *Cartoon drawings of an interesting possible arrangement of domains within the plasma-membrane H^+-ATPase of fungi. (A) Cytoplasmic portion, with the D segment (Gln161 to Ser283) removed to reveal a hypothetical catalytic cavity (stippled). (B) Hydrophobic helices, which would span the membrane, represented as cylinders in a slightly twisted bundle. In the intact molecule A would be attached to B by connections indicated in part E. (C and D) Cross sections of A and B, respectively.*

loop (42 aa's from H6 to H7), and a short C-terminus beyond H8 (40 aa's, $\mathbf{T_C}$).

In structure, overall shape, general physical properties, and nucleotide chemistry the fungal enzymes very closely resemble enzymes which pump Na^+ & K^+, K^+ & H^+, Ca^{++} & H^+, etc., through animal cell membranes. The resemblance is obvious despite many differences of the primary sequence: less than 25% amino acid identity between animal and fungal enzymes. Evidently, the same idea has been used to accomplish active transport of cations by plasma membranes of all living cells, and the differing cation specificities and coupling ratios have been incorporated into (as yet unidentified) details of the amino acid sequences. Evolution, then, has been economical. It has solved the commanding problem of osmotic regulation in animal cells using the same design of fuel cell as it used to balance a highly acid metabolism in plants and fungi.

Notes

The author is indebted to many people for assistance in preparing this essay and the original lecture. Special thanks are due to Ms. Barbara Gilson, of the Rockefeller University Archives, and Ms. Beth Carroll-Horrocks, of the American Philosophical Society, who provided biographical material on W. J. V. Osterhout; to Drs. Nina Allen, of the MBL and Wake Forest University, and N. Alan Walker, of the University of Sydney, who gave advice and beautiful illustrations of *Nitella* and *Chara*; to Drs. Dietrich Gradmann and Adam Bertl of the University of Göttingen, for potassium channel data; and to Drs. Colin Willmer, of the University of Stirling, and Ruth Satter[3], of the University of Connecticut, for illustrations of stomata and pulvinus.

LITERATURE CITED

Abe, T., and K. Oda. 1976. Resting and action potentials of excitable cells in the main pulvinus of *Mimosa pudica. Plant & Cell. Physiol.* **17**: 1343-1346.

Alcock, N. H. 1906. The action of anaesthetics on living tissues. I. The action on isolated nerve. *Proc. Roy. Soc. Lond. Ser B* **77**: 267-283.

Amberson, W. R. 1936. On the mechanism of the production of electromotive forces in living tissues. *Cold Spring Harbor Symp. Quant. Biol.* **4**: 53-59.

Barry, P. H. 1970. Volume flows and pressure changes during an action potential in cells of *Chara australis*. I. Experimental results. *J. Membr. Biol.* **3**: 313-334.

Bartley, W., and R. E. Davies. 1954. Active transport of ions by subcellular particles. *Biochem. J.* **57**: 37-49.

Bates, G. W., M. H. M. Goldsmith, and T. H. Goldsmith. 1982. Separation of tonoplast and plasma membrane potential and resistance in cells of oat coleoptiles. *J. Membr. Biol.* **66**: 15-23.

[3]Deceased August 3, 1989.

Bernstein, J. 1902. Untersuchungen zur Thermodynamik der bioelektrischen Ströme. *Pflügers Arch.* **92**: 521-562.

Bertl, A., and D. Gradmann. 1987. Current-voltage relationships of potassium channels in the plasmalemma of *Acetabularia. J. Membr. Biol.* **99:** 41-49.

Blatt, M. R. 1988. Potassium-dependent, bipolar gating of K^+ channels in guard cells. *J. Membr. Biol.* **102**: 235-246.

Brierley, G. P. 1976. The uptake and extrusion of monovalent cations by isolated heart mitochondria. *Mol. Cell. Biochem.* **10**: 41-62.

Burdon-Sanderson, J. 1873. Note on the electrical phenomena which accompany irritation of the leaf of *Dionaea muscipula. Proc. Roy. Soc. Lond.* **21**: 495-496.

Burdon-Sanderson, J. 1883. On the electromotive properties of the leaf of *Dionaea* in the excited and unexcited states. *Phil. Trans. R. Soc. Lond.* **173**: 1-55.

Cole, K. S. 1940. Permeability and impermeability of cell membranes for ions. *Cold Spring Harbor Symp. Quant. Biol.* **8**: 110-122.

Cole, K. S., and H. J. Curtis. 1938. Electric impedance of *Nitella* during activity. *J. Gen. Physiol.* **22**: 37-64.

Engelman, D. M., T. A. Steitz, and A. Goldman. 1986. Identifying nonpolar transbilayer helices in amino acid sequences of membrane proteins. *Ann. Rev. Biophys. Biophys. Chem.* **15**: 321-353.

Etherton, B., and N. Higinbotham. 1960. Transmembrane potential measurements of cells of higher plants as related to salt uptake. *Science* **131**: 409-410.

Findlay, G. P., and A. B. Hope. 1964. Ionic relations of cells of *Chara australis*. VII. The separate electrical characteristics of the plasmalemma and tonoplast. *Aust. J. Biol. Sci.* **17**: 62-77.

Gaffey, C. T., and L. J. Mullins. 1958. Ion fluxes during the action potential in *Chara*. *J. Physiol.* **144**: 505-524.

Ghislain, M., A. Schlesser, and A. Goffeau. 1987. Mutation of a conserved glycine residue modifies the vanadate sensitivity of the plasma membrane H^+-ATPase from *Schizosaccharomyces pombe*. *J. Biol. Chem.* **236:** 17549-17555.

Goldfarb, V., D. Sanders, and D. Gradmann. 1984. Reversal of electrogenic Cl^- pump in *Acetabularia* increases level and ^{32}P labelling of ATP. *J. Exp. Bot.* **35**: 645-658.

Goldman, D. E. 1943. Potential, impedance, and rectification in membranes. *J. Gen. Physiol.* **27**: 37-60.

Gorton, H. L. 1987. Water relations in pulvini from *Samanea saman*. *Plant Physiol.* **83**: 945-950.

Gradmann, D. 1975. Analog circuit of the *Acetabularia* membrane. *J. Membr. Biol.* **25**: 183-208.

Gradmann, D., U. -P. Hansen, W. S. Long, C. L. Slayman, and J. Warncke. 1978. Current-voltage relationships for the plasma membrane and its principal electrogenic pump in *Neurospora crassa*. I. Steady-state conditions. *J. Membr. Biol.* **39**: 333-367.

Graham, J., and R. W. Gerard. 1946. Membrane potentials and excitation of impaled single muscle fibers. *J. Gen. Physiol.* **28**: 99-116.

Guy, H. R. 1984. A structural model of the acetycholine receptor channel based on partition energy and helix packing calculations. *Biophys. J.* **45**: 249-261.

Hager, K. M., S. M. Mandala, J. W. Davenport, D. W. Speicher, E. J. Benz, Jr., and C. W. Slayman. 1986. Amino acid sequence of the plasma membrane ATPase of *Neurospora crassa*: deduction from genomic and cDNA sequences. *Proc. Nat. Acad. Sci. U.S.A.* **83**: 7693-7697.

Hamill, O. P., A. Marty, E. Neher, B. Sakmann, and F. J. Sigworth. 1981. Improved patch-clamp techniques for high-resolution current recording from cells and cell-free membrane patches. *Pflügers Arch.* **391**: 85-100.

Hedrich, R., and J. L. Schroeder. 1989. The physiology of ion channels and pumps in higher plant cells. *Ann. Rev. Pt. Physiol. Pt. Mol. Biol.* **40**: 539-569.

Hennessey, J. P., Jr., and G. A. Scarborough. 1988. Secondary structure of the *Neurospora crassa* plasma membrane H^+-ATPase as estimated by circular dichroism. *J. Biol. Chem.* **263:** 3123-3130.

Hermann, L. 1905. Beitrage zur Physiologie und Physik des Nerven. *Pflügers Arch.* **109**: 95-144.

Higinbotham, N., B. Etherton, and R. J. Foster. 1964. Effect of external K, NH_4, Na, Ca, Mg, and H ions on the cell transmembrane electropotential of *Avena coleoptile*. *Plant Physiol.* **39**: 196-203.

Hodgkin, A. L., and A. F. Huxley. 1952. The components of the membrane conductance in the giant axon of *Loligo*. *J. Physiol.* **116**: 473-496.

Hope, A. B., and N. A. Walker. 1975. *The Physiology of Giant Algal Cells.* Cambridge University Press, Cambridge. 201 pp.

Huganir, R. L. 1988. Regulation of the nicotinic acetylcholine receptor channel protein by phosphorylation. *Curr. Top. Membr. Transp.* **33**:147-163.

Hüsken, D., E. Steudle, and U. Zimmermann. 1978. Pressure probe technique for measuring water relations of cells in higher plants. *Plant Physiol.* **61**:158-163.

Laver, D. R., and N. A. Walker. 1987. Steady-state voltage-dependent gating and conduction kinetics of single K^+ channels in the membrane of cytoplasmic drops of *Chara australis*. *J. Membr. Biol.* **100** :31-42.

Mailman, D. S., and L. J. Mullins. 1965. The electrical measurement of chloride fluxes in *Nitella. Aust. J. Biol. Sci.* **19**: 385-398.

Mandala, S. M., and C. W. Slayman. 1988. Identification of tryptic cleavage sites for two conformational states of the *Neurospora* plasma membrane H^+-ATPase. *J. Biol. Chem.* **263**: 15122-15128.

Moore, J. W., and K. S. Cole. 1960. Resting and action potentials of the squid giant axon *in vivo*. *J. Gen. Physiol.* **43**:961-970.

Moran, N., G. Ehrenstein, K. Iwasa, C. Mischke, C. Bare, and R. L. Satter. 1988. Potassium channels in motor cells of *Samanea saman. Plant Physiol.* **88**: 643-648.

Mummert, H., and D. Gradmann. 1976. Voltage-dependent potassium fluxes and the significance of action potentials in *Acetabularia. Biochim. Biophys. Acta* **443:** 443-450.

Neher, E., and B. Sakmann. 1976. Single-channel currents recorded from membrane of denervated frog muscle fibres. *Nature* **260:** 779-802.

Nernst, W. 1890. Die elektromotorische Wirksamkeit der Ionen. *Z. Physik. Chem.* **4:** 129-181.

Osterhout, W. J. V. 1905. *Experiments with Plants*. MacMillan Co., London. 492 pp.

Osterhout, W. J. V. 1922. *Injury, Recovery, and Death, in Relation to Conductivity and Permeability*. J.B. Lippincott Co., Philadelphia, PA. 259 pp.

Osterhout, W. J. V. 1927. Some aspects of bioelectrical phenomena. *J. Gen. Physiol.* **11**: 83-99.

Osterhout, W. J. V. 1931. Physiological studies of single plant cells. *Biol. Rev.* **6**: 369-411.

Osterhout, W. J. V. 1934. Nature of the action current in *Nitella. J. Gen. Physiol.* **18:** 215-227.

Osterhout, W. J. V. 1946. Nature of the action current in *Nitella.* VI. Simple and complex action patterns. *J. Gen. Physiol.* **30**: 47-59.

Osterhout, W. J. V., and E. S. Harris. 1927. Protoplasmic asymmetry in *Nitella* as shown by bioelectric measurements. *J. Gen. Physiol.* **11:** 391-406.

Osterhout, W. J. V., and S. E Hill. 1936. Some ways to control bioelectrical behavior. *Cold Spring Harbor Symp. Quant. Biol.* **4**: 43-52.

Paszewski, A., and T. Zawadzki. 1974. Action potentials in *Lupinus angustifolius* L. shoots. II. Determination of the strength-duration relation and the all-or-nothing law. *J. Exp. Bot.* **25**: 1097-1103.

Peebles, M. J., F. V. Mercer, and T. C. Chambers. 1964. Studies on the comparative physiology of *Chara australis. Aust. J. Biol. Sci.* **17**: 49-61.

Penny, M. G., and D. J. F. Bowling. 1974. A study of potassium gradients in the epidermis of intact leaves of *Commelina communis* L. in relation to stomatal opening. *Planta* **119:** 17-25.

Pickard, B. G. 1971. Action potentials resulting from mechanical stimulation of pea epicotyls. *Planta* **97:**106-115.

Planck, M. 1890. Ueber die Erregung von Electricität und Wärme in Electrolyten. *Ann. Phys. Chem. (NF)* **39**:161-186.

Raschke, K., R. Hedrich, U. Reckmann, and J. I. Schroeder. 1988. Exploring biophysical and biochemical components of the osmotic motor that drives stomatal movement. *Botanica Acta* **101**: 283-294.

Satter, R. L. 1979. Leaf movements and tendril curling. Pp. 442-484 in *Encyclopedia of Plant Physiology*, New Series Vol. 7: *Plant Movements*. W. Haupt and M.E. Feinleib, eds. Springer-Verlag, Berlin.

Satter, R. L., R. C. Garber, L. Khairallah, and Y. -S. Cheng. 1982. Elemental analysis of freeze-dried thin sections of *Samanea* motor organs: barriers to ion diffusion through the apoplast. *J. Cell Biol.* **95:** 893-902.

Satter, R. L., G. T. Geballe, P. B. Applewhite, and A. W. Galston. 1974. Potassium flux and leaf movement in *Samanea saman. J. Gen. Physiol.* **64:** 413-430.

Schroeder, J. I. 1988. K^+ transport properties of K^+ channels in the plasma membrane of *Vicia faba* guard cells. *J. Gen. Physiol.* **92**: 667-683.

Schroeder, J. I., and S. Hagiwara. 1989. Cytosolic calcium regulates ion channels in the plasma membrane of *Vicia faba* guard cells. *Nature* **338**: 427-430.

Schroeder, J. I., R. Hedrich, and J. M. Fernandez. 1984. Potassium-selective single channels in guard cell protoplasts of *Vicia faba. Nature* **312**: 361-362.

Schroeder, J. I., K. Raschke, and E. Neher. 1987. Voltage dependence of K^+ channels in guard-cell protoplasts. *Proc. Nat. Acad. Sci. U.S.A* . **84**: 4108-4112.

Serrano, R., M. C. Kielland-Brandt, and G. R. Fink. 1986. Yeast plasma membrane ATPase is essential for growth and has homology with (Na^++ K^+), K^+- and Ca^{2+}-ATPases. *Nature* **319**: 689-693.

Sibaoka, T. 1966. Action potentials in plant organs. *Symp. Soc. Exp. Biol.* **20**: 49-73.

Skou, J. C. 1957. The influence of some cations on an adenosine triphosphatase from peripheral nerves. *Biochim. Biophys. Acta* **23**: 394-401.

Slayman, C. L. 1965. Electrical properties of *Neurospora crassa*: respiration and the intracellular potential. *J. Gen. Physiol.* **49**: 93-116.

Slayman, C. L., P. Kaminski, and D. Stetson. 1989. Structure and function of fungal plasma-membrane ATPases. Pp. 295-312 in

Biochemistry of Cell Walls and Membranes in Fungi, P. J. Kuhn *et al.*, eds. Springer-Verlag, Berlin.

Slayman, C. L., W. S. Long, and D. Gradmann. 1976. "Action potentials" in *Neurospora crassa,* a mycelial fungus. *Biochim. Biophys. Acta* **426**: 732-744.

Slayman, C. L., C. Y. -H. Lu, and L. Shane. 1970. Correlated changes in membrane potential and ATP concentrations in *Neurospora. Nature* **226**: 274-276.

Slayman, C.L., and C.W. Slayman. 1962. Measurement of membrane potentials in *Neurospora. Science* **136**: 876-877.

Snow, R. 1924. Conduction of excitation in stem and leaf of *Mimosa pudica. Proc. R. Soc. Lond. B.* **96**: 349-374.

Spanswick, R. M., J. Stolarek, and E. J. Williams. 1967. The membrane potential of *Nitella translucens. J. Exp. Bot.* **18**: 1-16.

Steinbach, E. 1896. Ueber die electromotorischen Ersheinungen an Hautsinnesnerven bei adaequater Reizung. *Pflügers Arch.* **63**: 495-520.

Taylor, C. V., and D. M. Whitaker. 1925. A measurable potential difference between the cell interior and outside medium. *Carnegie Inst. Wash. Year Book.* **25**: 248-249.

Walker, N. A. 1955. Microelectrode experiments on *Nitella. Aust. J. Biol. Sci.* **8**: 476-489.

Weidmann, S. 1957. Resting and action potentials of cardiac muscle. *Ann. NY Acad. Sci.* **65**:663-678.

Wendler, S., U. Zimmermann, and F. -W. Bentrup. 1983. Relationship between cell turgor pressure, electrical membrane potential, and chloride efflux in *Acetabularia mediterranea. J. Membr. Biol.* **72**: 75-84.

Williams, S. E., and B. G. Pickard. 1972. Properties of action potentials in *Drosera* tentacles. *Planta* **103**: 333-340.

Willmer, C. M. 1983. *Stomata*. Longman Group, Ltd., London. 166 pp.

Kenneth S. Cole
(1900 - 1984)

Alan L. Hodgkin
(1914 -   )

Sodium and Potassium Channels and Propagation of the Nerve Impulse: A Tribute to Cole and Hodgkin

Introduction by Franklin F. Offner

In the mid 1930s, how action potentials are generated, and how they are conducted, was still a mystery—almost mystical. In this atmosphere, **K. C. Cole**'s approach, a physical measurement of the electrical parameters of the membranes (Cole, 1933; Cole and Curtis, 1936), stood out as a rational, scientific approach to the problem. This is what attracted me to him and his work.

The electrical parameters of membranes in fact proved to be of fundamental importance to the understanding of membrane activity. Bernstein (1902) had postulated activity to be the result of the membrane's loss, in some unknown way, of its potassium selectivity. A simple mathematical model (Offner, 1939; Offner *et al.*, 1940) showed that if this was the result of an increased leakage current, the change could be propagated along the axon as an action potential, and the increased leakage should be seen as a lowering of the membrane resistance. This is just what Cole and Curtis found in the squid axon in 1938 (Cole and Curtis, 1938). About the same time, **A. L. Hodgkin** demonstrated that the neighboring region of the axon will then be stimulated by electronic spread from the excited region, as required for conduction.

So the nature of the action potential started to become a little less mysterious. But now a new mystery appeared, when in 1939 both Cole and Curtis, and Hodgkin and Huxley, recorded the action potential in squid axon with an internal electrode, and found it to be almost twice as large as the resting potential. This didn't fit the model; where did the extra potential come from? While Cole found some impedance effects he thought might account for it, Hodgkin hypothesized that it was the result of a sodium current, but was unable to prove it experimentally, until Cole made another fundamental advance. This was after the war, when Cole had moved to the University of Chicago. There, as an outgrowth of the current clamp developed by Marmont (1949), he developed the voltage clamp method for recording the ionic currents produced by an axon, a technique that has since been the basis for most experimental advances in this field. Hodgkin visited Cole that year, to become familiar with the voltage clamp (Cole, 1968); and returning to England, he and Huxley used this new technique to perform the remarkable series of experiments on squid axon, from which they were able to describe in a concise manner the basic ionic phenomena which take place in a wide spectrum of membranes (Hodgkin and Huxley, 1952). All of us in the field have since built on this cooperation between Cole and Hodgkin, and their colleagues.

I personally probably had a closer, and certainly a longer, relationship with Kenneth Cole, than I have had with anyone else in the field. It started in 1936, when I first visited him at Columbia, while I was working in Ralph Gerard's laboratory at the University of Chicago; and thereafter, whenever I could. Chicago and New York City were too far apart for frequent visits at that time, but I recall visiting him in 1937, to ask him about my idea of measuring the predicted impedance change in *Nitella* during activity. He told me that he and Curtis had just finished doing just that (Cole, 1968)! Later, I was able to visit Kacy more often at Columbia, when I was working in New York City. This was when their impedance measurements were giving some rather strange results, which Kacy referred to as "inductive reactance." We had many theoretical discussions, our viewpoints being somewhat different. Later, when Kacy was at the NIH, and part of each year at Berkeley, I would visit him whenever I could, spending some very pleasant and useful evenings together. He would keep me up to date on the latest work in the field, and help keep me straight in my work. In the later years, as the last survivors of an earlier era, we would spend a lot of time together at the Biophysical Society meetings, going over the old times. He was tremendously saddened by the death of his wife; and towards the end, his illness reduced our get-togethers to talks on the phone. His passing leaves me saddened, and with a sense of deprivation.

Clay Armstrong is Professor of Physiology at the University of Pennsylvania, and is one of the leaders in the field, of a later generation. He received an MD from Washington University, where modern axonology started some 60 years ago, with such great names as Erlanger, Gasser, and George Bishop. He joined Kacy's laboratory in NIH in 1961, and made his first appearance in Woods Hole shortly thereafter. He became enamored with the problem of nerve conduction at that time, and except for a wonderfully instructive two years with Huxley at University College, London, he has worked on nerve conduction and the properties of ionic channels almost exclusively since that first exposure. Using various probes, one of Clay's first contributions was the use of TEA and its derivatives to selectively block potassium channels (Armstrong and Binstock, 1965), which is now in universal use. Similarly, his work with pronase (Armstrong *et al.*, 1973) provided a major step in understanding the mechanism of sodium inactivation. Perhaps Clay is best known for his work with Bezanilla in recording the "gating currents" associated with the opening and closing of ionic channels in membranes (Armstrong, 1981). The examination of these currents has provided information useful in examining the validity of theories of membrane function.

Literature Cited

Armstrong, C. M. 1981. Sodium channels and gating currents. *Physiol. Rev.* **61:** 644-683.

Armstrong, C. M., and L. Binstock. 1965. Anomalous rectification in the squid giant axon injected with tetraethylammonium chloride. *J. Gen. Physiol.* **48:** 859-872.

Armstrong, C. M., F. Bezanilla, and E. Rojas. 1973. Destruction of sodium conductance inactivation in squid axon perfused with Pronase. *J. Gen. Physiol.* **62:** 375.

Bernstein, J. 1902. Untersuchungen zur Thermodynamik der bioelektrischen Ströme I (Erster Theil). *Arch. ges. Physiol.* **92:** 521-562.

Cole, K.S. 1933. Electrical conductance of biological systems. *Cold Spring Harbor Symp. Quant. Biol.* **1:** 1-11.

Cole, K. S. 1968. *Membranes, Ions, and Impulses.* University of California Press. Pp. 267-269.

Cole, K. S., and H. J. Curtis, 1936. Electrical impedance of nerve and muscle. *Cold Spring Harbor Symp. Quant. Biol.* **4:** 1-40.

Cole, K. S., and H. J. Curtis, 1938. Electrical impedance of nerve during activity. *Nature* **142:** 209.

Hodgkin, A. L., and A. F. Huxley. 1952. A quantitative description of membrane current and its application to conduction and excitation in nerve. *J. Physiol.* **117:** 500-544.

Marmont, G. 1949. Studies of the axon membrane; I. A new method. *J. Cell. Comp. Physiol.* **34:** 351-382.

Offner, F. F. 1939. Current theory of nervous conduction. *Am. J. Physiol.* **126:** 594.

Offner, F. F., A. Weinberg, and G. Young, 1940. Nerve conduction theory: some mathematical consequences of Bernstein's model. *Bull. Math. Biophys.* **2:** 89-103.

Sodium and Potassium Channels and Propagation of the Nerve Impulse: A Tribute to Cole and Hodgkin

Clay M. Armstrong

University of Pennsylvania Medical School

ELECTRICAL ACTIVITY IN THE NERVOUS SYSTEM is such a fundamental part of our being that it is astonishing to remember that our understanding of it was firmly established only in this century. K. S. Cole and A. L. Hodgkin are two of the great architects of our understanding. Cole spent his summers at the MBL during most of his active career, and it was at MBL that he began his famous experiments on the squid giant axon. It was in Cole's MBL laboratory in 1938 that Hodgkin first worked on this remarkable fiber. The giant axon (Fig. 1) was originally described by Williams (1910), working at MBL, and rediscovered by J. Z. Young (1936), who brought it to the attention of Cole and others. It proved to be the ideal experimental preparation, having a simple, cylindrical geometry and a very large size: twenty times the diameter of the largest human nerve fiber. Its size was a major advantage, because it allowed the curious investigator to introduce electrodes inside, and, for the first time, directly measure the voltage changes during an action potential.

Cole, Hodgkin, and their brilliant coworkers, including A. F. Huxley and Bernard Katz, exploited the giant axon to lay the basis for understanding of electrical activity in the nervous system. The framework they created is breathtakingly comprehensive, one of the major accomplishments of biological science in the twentieth century.

To appreciate the significance of this work, one need only think of the variety of ways in which electrical activity is important. Nerve cells are specialized for signaling, and can transmit coded messages over long distances within the body in milliseconds. To mention only a few other examples, electrical signals determine the timing of the heart beat, control activation of mechanical activity in muscle fibers, time the contractions in some types of smooth muscle, and play an essential role in the secretion of neurotransmitters and hormones.

These sophisticated mechanisms clearly did not evolve quickly. This chapter offers a short summary of how cellular electrical properties may have evolved over the course of a billion or more years, and concurrently traces the evolution of our understanding of these properties over the much briefer span of about one hundred years. The historical aspects rely heavily on the account by Cole (1968), and the excellent book by Bertil Hille (1984).

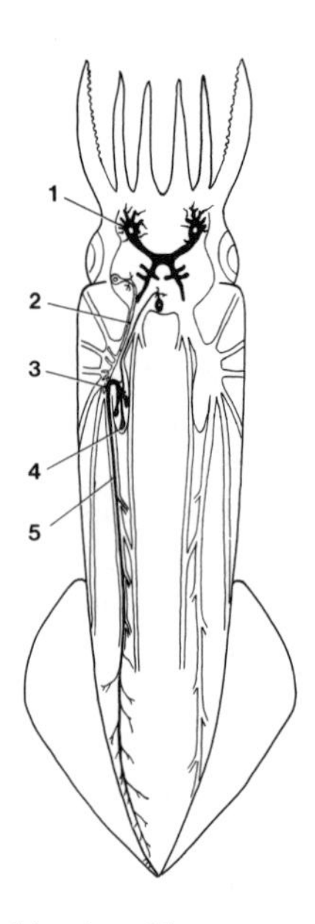

Figure 1. *A squid and its giant fiber system. Nervous impulses from the brain are transmitted to the squid's mantle (body) by the first (1) and second (2) order giant axons. The second order giant axon synapses (3) with the third order giant axon (5), which is formed by the fusion of the small axonal processes of many neurons (4) in the stellate ganglion. Modified from J. Z. Young (1936).*

Why Cells Need Membrane Voltage

No living cell that I am aware of lacks a membrane potential. A cell can be defined for our purposes as having an inside and an outside, separated by a surface membrane. The surface membrane is a selectively permeable barrier that allows the cell to sequester inside substances that are vital to its existence. From the cell's point of view, the external medium is infinite in volume. It is the source of all raw materials needed by the cell, and is the final repository for all waste products. An axon, which will figure heavily in the discussion, is a long, thin, cylindrical projection from the cell, with a specialized ending that secretes neurotransmitters.

Because all cells have a membrane voltage, with a resting value of typically -60 to -90 mV, it seems clear that the voltage must play an essential role within the cell. Further, voltage must be essential not only to nerve and muscle cells, but also to liver cells, plant cells, and bacteria. How are these membrane potentials established, and what role do they play?

Perhaps surprisingly, voltages occur whenever there is contact between two solutions that differ in their content of dissolved electrolytes. This was precisely quantitated by Nernst (1888). Bernstein (1902) used Nernst's results to formulate a stunningly perceptive guess relating ions and voltages in cells. He hypothesized that cells are surrounded by a membrane that is selectively permeable to potassium ion. Because the potassium ion concentration inside the cell is higher than outside, the result, according to Nernst, should be a membrane voltage that is negative inside. Further, Bernstein proposed that changes in the membrane voltage were the result of a permeability increase that led to a loss of selective K permeability, and thus a temporary disappearance of the voltage. In a sense, there was little to add.

What essential role do voltages play in the life of all cells, even bacteria? There are at least two answers to this question. First, it is now accepted that membrane voltage serves as an intermediate storage form in the production of chemical energy in bacteria and mitochondria. According to Mitchell's hypothesis (1961), energy from the oxidative metabolism of sugar is used to pump protons from the interior of a bacterium, the result being the development of a low proton concentration inside the bacterium, and a large negative voltage. The tendency of positively charged protons to flow downhill energetically to the region of negative voltage and proton scarcity, constitutes an energy store that is tapped by the bacterium to produce ATP—the chemical gasoline of the cell. Protons are allowed to reenter the bacterium, like water flowing downhill, through a specialized 'channel.' Protons that flow through this channel turn an ATP-producing turbine. The evolution of this elaborate machinery, which is still not known in molecular detail, resulted in a large increase in the amount of ATP that could be generated from the metabolism of a sugar molecule.

Membrane voltage has a second essential role in animal cells, in which the cell membrane is not supported by an external skeleton: it helps control the volume of the cell. Cell volume control is an interesting story that is not widely appreciated outside of a rather small circle of physiologists. How does a cell maintain its volume? One's first thought is, that like any vessel, the cell has rigid walls that contain a certain volume, just as a glass will contain a certain volume of liquid. Although this picture is plausible for a plant cell that has an external skeleton, it is not adequate for an animal cell. Our cells are enclosed by a surface membrane that has about the same tensile strength as a soap bubble.

The problem is compounded by the cell's need to save essential substances in its interior. A store of amino acids, for example, raises the internal osmolarity (the total concentration of dissolved substances), and reciprocally, lowers the concentration of water. This creates a tendency for water to enter the cell, following its concentration gradient. The surface membrane is not an effective barrier to water movement, and water flows in rather freely, causing the cell to swell. The forces involved are very large, and the fragile surface membrane offers no significant resistance. Swelling continues, in theory, until the sequestered substances are infinitely dilute, and volume infinitely large. The result would be the lysis of the cell, as occurs when red blood cells are suspended in distilled water: they quickly swell and break. Clearly, then, a cell with an unsupported membrane can survive only if the osmolarity is the same inside and outside. But how is this compatible with the cell's need to hoard materials like sugars and amino acids?

The answer is that the cell must eject some substance that it does not need, and the chosen substance is chloride ion. Chloride ions have a negative charge, and are driven out of the cell by the negativity inside the cell, thus lowering the concentration of dissolved particles inside the cell and the osmolarity. Although the details of cell volume control are complex, it is enough for our purposes to say that just enough chloride is driven out to compensate for the osmolarity of sugars and amino acids hoarded within the cell.

Driving out chloride ion requires energy, and this energy is supplied in the form of ATP. Specifically, ATP (made predominantly by the mitochondria) is consumed by a 'pump' molecule (called the Na/K pump) in the surface membrane, which drives Na out of the cell in exchange for K ion. The pump thus makes the K ion concentration inside the cell higher than outside, as Bernstein had postulated, and keeps the internal Na concentration low. The combination of a high internal K concentration, and a membrane that is selectively permeable to K, creates a membrane potential that is negative inside.

This mechanism is wonderfully indirect, and complex because many pieces have to fit together. To recapitulate, (i) mitochondria make ATP by metabolizing sugar products. The transmembrane voltage of the mitochondrion is an essential element of the process. (ii) The Na/K pump of the cells surface membrane uses ATP to exchange Na for K internally, making the K concentration inside the cell high (approximately 150 m*M* inside, against 5 m*M* K outside). (iii) The high internal K concentration combined with a membrane that is selectively permeable to K ion makes a membrane voltage that is negative inside. (iv) The negative membrane voltage drives out chloride ions, making osmotic room for the substances that the cell wants to save in its interior.

Other Uses of Membrane Voltage

Thus cells require membrane voltages for at least two very basic purposes: energy production and volume control. It follows that a membrane voltage must have been a property of even very early cells. Over the billions of years since, many sophisticated uses of membrane voltage have been evolved, and one of the most extraordinary is the adaptation of electricity for signaling purposes. A complex organism needs a communication system to coordinate its various parts, and organisms use three basic mechanisms. One is diffusion, which is effective for short distances. The second coordinating mechanism is the introduction of signaling substances, *e.g.*, hormones into the circulatory system. The third is electrical signaling. It is instructive to compare the time that these three systems would require to deliver a message from head to toe. For electrical signaling the time is measured in milliseconds; for the circulatory system in seconds; and for diffusion in centuries! The advantages of electrical signaling for an organism as large as a human are thus clear.

Electrical signaling in cables, *e.g.*, the trans-Atlantic cable from England to the United States, was well understood in the nineteenth century (Kelvin, 1855), and these insights had spread to biology at least by the turn century (Cremer, 1900; Herman, 1905). The basic idea is that an electrical signal initiated at one end of a trans-Atlantic cable spreads out along the cable, and an electrical signal starting at one end of an axon would do the same. In both cases, the signal current flows through the conductor of the cable (the copper wire, or the electrolyte solution which is the cell cytoplasm), and is prevented from escaping to the outside (the ocean, or the external medium) by an insulator (the insulation surrounding the cable, or the surface membrane). The cell membrane, it was recognized, is far from a perfect insulator, and, in the resting state, allows enough K ions to pass to establish the resting voltage, as Bernstein said.

Although these hypotheses pointed the way, there was an enormous amount to learn. How far would the signal spread in a cellular cable? And, as it dissipated and lost sharpness along the length of the cable (as, distressingly, did the signal in the first trans-Atlantic cable) how was the signal boosted back to full strength? The action potential occurs only when voltage has crossed a threshold level. How could the threshold be explained? These questions required a systematic approach, which continued through the first half of the century, and were fully solved in 1952 with the publication of the work of Hodgkin and Huxley.

The first step was to get realistic estimates of the electrical properties of the cytoplasm and the membrane, and in this respect the squid giant axon was, thanks to its large size and simple geometry, a perfect preparation. Cole and his coworkers, through ingenious experiments, soon confirmed that the membrane is sufficiently high in resistance to be called an insulator, and the cytoplasm is low enough in resistance to be called a conductor. Thus, an axon had the fundamental characteristics of a cable. Further, they showed, in a great and justly famous experiment, that the membrane resistance drops during an action potential, and recovers quickly after it is over (Fig. 2). These results gave important support to Bernstein's hypothesis that during activity there was an increase in permeability (drop in membrane resistance).

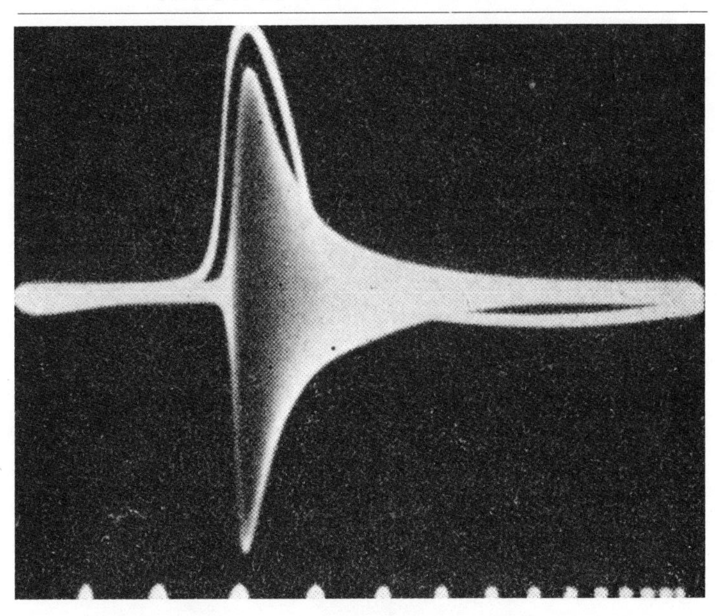

Figure 2. *Membrane voltage (V_m) and impedance (Z_m) of an axon membrane during an action potential. The Z_m envelope is the output of a Wheatstone bridge, and indicates a large drop in membrane resistance during the action potential. The action potential lasts about one millisecond. From Cole and Curtis, 1939.*

The next major finding was a complete surprise. Measurements of electrical activity so far had been confined to the ones that could be made with extracellular electrodes. At about the same time, Hodgkin and Huxley in England and Curtis and Cole, decided it would be feasible to put an electrode inside to record the transmembrane voltage. (Fig. 3 is a photomicrograph of this audacious procedure, showing an axon containing two spirally wound electrodes.) As Cole tells it, he was not very excited by the prospect, and thought that seeing an action potential upside-down would not be very interesting (Cole, 1968). The result, however, was of enormous interest, and demonstrated that, rather than approaching zero, as Bernstein's theory predicted, the membrane voltage reversed sign, and became (internally) positive. This finding became the cornerstone of the sodium

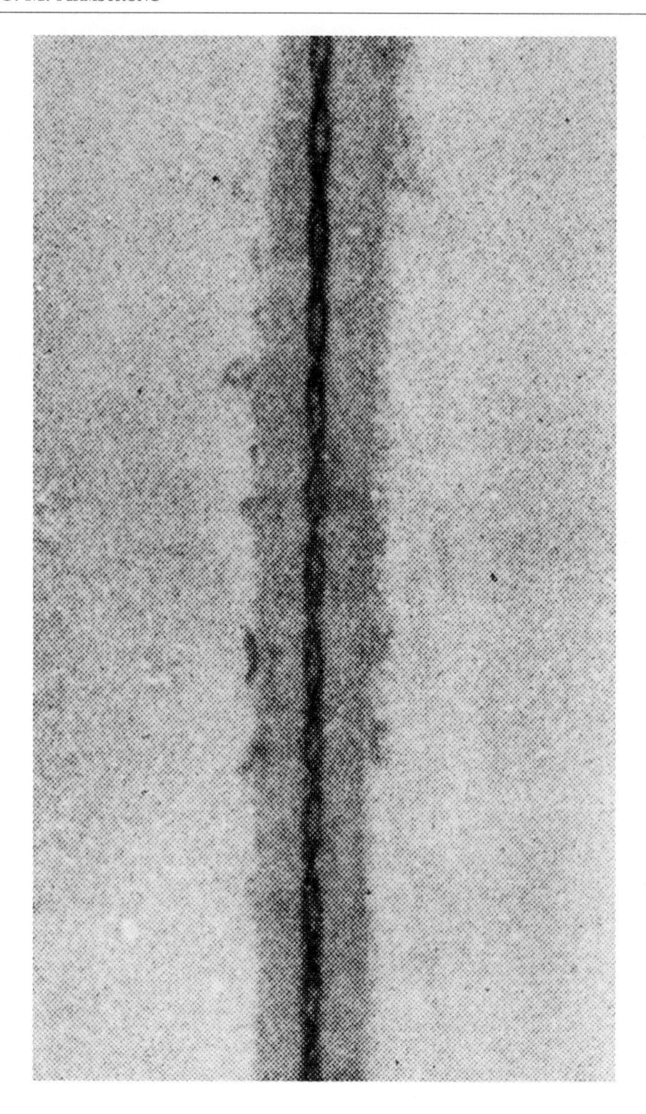

Figure 3. *A photomicrograph of an isolated squid giant axon into which an axial electrode has been placed. The axial electrode consists of two helically wound wires, one for measuring membrane voltage, the other for passing current through the membrane. The axon is approximately 1 mm in diameter. From Hodgkin, Huxley, and Katz (1952; Fig. 5A).*

theory. Overton had suggested that Na ion plays an important part in excitability, and Hodgkin and Katz (1949) soon found it was essential. During activity, the membrane did not simply lose its selective K permeability as Bernstein had hypothesized, but changed from K selective to Na selective. The finding agreed well with the resistance measurements of Cole and Curtis (1939) if one imagined there was a sudden increase in permeability to Na ion. Cole generously comments that when Hodgkin first explained the sodium theory to him, he was far from understanding either the theory or its implications.

Was it reasonable to think of a major permeability change that occurred in less than a millisecond? How could one investigate this startling proposition? The necessary tools, the 'space clamp' and the voltage clamp, were developed in Cole's laboratory. During activity, the voltage along an axon changes continuously, introducing a complication that would better be avoided. Marmont, in Cole's lab, solved this problem by, in effect, shorting out the resistance of the cytoplasm, by inserting an internal wire all the length of the axon segment. This 'space clamp' meant that all parts of the axon would be placed in lock step, giving the experimenter in essence a single large patch of membrane. Equally essential was the voltage clamp, which made it possible to clamp the membrane voltage at any desired level, and prevent it from changing continuously. The voltage clamp had not only technical advantages (it eliminated the capacitive component of current) but it also made the results much easier to think about. The design of the voltage clamp, which operated on the principle of negative feedback, was possible because of advances in control theory, before and during the Second World War.

Cole used space and voltage clamp techniques to obtain the first records, which showed the now-familiar pattern of an inward current for about a millisecond, followed by an outward current. These currents were precisely what was needed to explain the upstroke (inward current) and the downstroke (outward current) of the action potential. But what lay behind these currents?

Hodgkin and his colleagues, Katz and Huxley, phrased this question in exactly the right way: what ions carry the inward and the outward current? To what ions was the membrane selectively permeable during the upstroke, and the downstroke of the action potential? So phrased, the answer was inescapable. Upon activation, the membrane changed from a resting state of low but predominantly K permeability, as Bernstein had said, to a state of high Na permeability. During the downstroke, Na permeability disappeared, and was replaced by a state of high K permeability. Within milliseconds after the end of the action potential, the elevated K permeability dropped to the resting level. But what triggered the rise in Na, and then K permeability? Why did the Na permeability go away after a millisecond or so? What physical principles could underlie this bewildering behavior? Was a chemical like acetylcholine somehow involved, as postulated by Nachmannsohn (1966)?

Hodgkin and Huxley used a much improved version of the voltage clamp to put all these pieces together, to form a framework that will stand forever. First, they postulated two separate sets of carriers, one for Na and one for K ions. The 'carriers' eventually became the Na and K channel proteins, as described later. The rise in Na permeability was mediated by activation of the Na carriers. This occurred whenever the membrane potential, beginning at a resting level of -60 or -70 mV, was driven positive to -50 mV. After about a millisecond at positive voltage, they found that the Na carriers spontaneously 'inactivated.' The overall effect was to produce a transient increase in the Na permeability. They described this behavior quantitatively in terms of two mathematical factors that governed the activity of the carriers, m (activation) and h (inactivation).

The potassium carriers had similar behavior, but the 'unleashing' after a positive change of membrane voltage was relatively slow. Thus a few Na carriers came into action first, and because there is more Na inside than out, they carried Na inward. This caused a positive change of membrane voltage, which unleashed more Na carriers, creating an element of positive feedback between voltage and number of activated carriers. This feedback caused the membrane voltage to rise very rapidly. After about a millisecond the Na carriers inactivated, and at about the same time the K carriers went into action. Net K movement was outward, which drove the internal voltage back toward the resting level, by letting positively charged K ions escape from the axon interior. This complex behavior is summarized in Figure 4.

An essential physical insight of the Hodgkin and Huxley formulation is that the carriers are controlled directly by the membrane voltage, with no intervention from chemicals such as acetylcholine. They imagined that the (unknown) structures that carried Na ions across the membrane were controlled by a 'component of the mem-

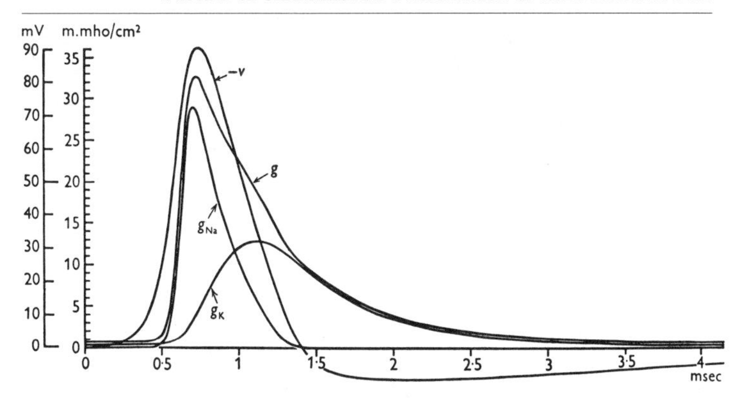

Figure 4. *Hodgkin and Huxley's reconstruction of the action potential. The membrane voltage (V) is driven up by the activation of sodium carriers (g_{Na}) and restored to its resting level by the activation of potassium carriers (g_K). The sum of g_{Na} and g_K is given by g, which resembles the impedance trace obtained by Cole and Curtis (Fig. 2). From Hodgkin and Huxley (1952; Fig. 17).*

brane that has a large charge or dipole moment.' This 'component of the membrane' moved or reoriented in response to a change of the membrane field, according to the usual laws of electricity, unleashing the Na carriers.

By a set of difficult and sophisticated calculations, which have given many generations of students much pause for thought, they showed that their 'model membrane' had the appropriate threshold, and predicted an action potential that is of the correct time course. Further, it predicted a propagating action potential. That is, if the action potential were stimulated at one end of an axon, it would propagate to the other end without decrement, and the calculated and experimentally observed velocities were the same. During propagation, depolarization spread ahead of the active region, and triggered the activation of the sodium carriers. 'Threshold' was reached when the inward current through the sodium carriers equaled or outweighed the K current carried by the still-dormant K carriers.

The Hodgkin and Huxley scheme provided all of the documentation necessary to confirm the essential correctness of earlier insights. In addition, their work provided a detailed description of something totally new, the Na and K carriers, which emerged as strong theoretical entities, the missing links in the story of action potential initiation and propagation. Although subsequent work has modified and extended their findings in some respects, one can confidently predict that their framework will be immortal, as will be the insightful but more general proposals of Bernstein.

To return to the trans-Atlantic cable analogy, one can now say that action potential propagation has two essential elements. The first is the cable itself, which, in the case of the axon, is of particularly poor quality, because it has a poor core conductor (the cytoplasm) surrounded by very leaky insulation (the membrane). The result is that an unboosted signal dies off rapidly as a function of distance along the axon, and must be returned to full strength by an amplification mechanism. In the trans-Atlantic cable, the amplifiers that detect the signal and reinforce it are spaced miles apart. In a myelinated axon, they are 1 or 2 mm apart (at the nodes of Ranvier), and they are continuously distributed in unmyelinated axons. The amplifiers are the Na channels, which detect the voltage change conducted by the cable from activity upstream, and reinforce it, by opening and driving the internal voltage positive.

With regard to the Na and K carriers themselves, there was still much to learn. The first question was whether one need think of two channels, or a single one that changes its selectivity from Na to K. This was a lively debate, fueled by the fact that inactivation of the Na channel had about the same time course as the activation of K channels. One could thus imagine a single channel that first conducted sodium, and then changed into a K channel (Mullins, 1959). Many interesting discoveries were made in attempting to answer this question. A very potent toxin, tetrodotoxin, was isolated from the puffer fish, and found to block Na channels without affecting K channels (Narahashi *et al.*, 1964). Tetraethlyammonium ion proved to be a good and rather selective blocker of K channels. Neither of these lines of evidence, although interesting in themselves, were completely convincing evidence for the existence of two channels. Mullins was at last convinced by the finding that inactivation of the Na channel could be destroyed by the proteolytic mixture pronase, without interfering with the increase of K permeability (Armstrong *et al.*, 1973). This meant that Na channels and K channels were conducting simultaneously, and must be separate structures.

The next question was how ions actually penetrate the membrane. Much evidence was accumulated to show that ions moved very rapidly through on the 'carriers,' and that it was simplest to think in terms of a continuous channel through the membrane. The ultimate proof came from direct measurement of the current through a single Na carrier, using the patch clamp techinque. This proved that ions move at the staggering rate of more than 10^7/s, much faster than can be explained by a carrier mechanism. Thus the Na carrier was in reality a sodium channel.

A fascinating question concerned selectivity: how could a channel be made selective for either Na or K ion? There has been considerable throretical progress on this question, and the best hypothesis to my mind is based on the close fit hypothesis of Mullins (1959). A dehydrated Na ion is smaller than a dehydrated K ion, and one can, in theory, explain exclusion of K ions from Na channels by a sieving effect: the K ion is too large to enter. But how can a Na ion be excluded from the K channel? Mullins postulated, reasonably, that the Na ion must not simply fit into the channel, but, for energetic reasons, must be a close fit. A clear resolution of the selectivity question is hindered by the relatively small selectivity ratio of both channels, about 100:1 (K/Na) for the K channel, and approximately 10:1 (Na/K) for the Na channel. From the energetic point of view, these numbers can be explained by postulating barriers to channel entry that differ, for the two ions, by only a fraction of the energy of a hydrogen bond. The factors responsible for such small energy differences may be difficult to identify even when the position of every atom in Na and K channels is known.

To me the most interesting question regarding Na and K channels has always been the molecular machinery that confers on Na and K channels their extreme sensitivity to voltage. As noted, Hodgkin and Huxley postulated that the Na carrier was activated after the movement or rotation of a controlling charge or dipole within the membrane. This led directly to the prediction of gating current: under favorable circumstances, one should be able to see the current generated by the movement of the controlling charge directly. Actual measurement of gating current (Fig. 5) was accomplished by Francisco Bezanilla and myself (Armstrong and Bezanilla, 1973), and quickly confirmed (Keynes and Rojas, 1974).

Gating current gave much information about the Na channel that could not be obtained in any other way. One contribution was to understanding the mechanism of inactivation, the spontaneous decline in sodium permeability that occurred shortly after the channels opened. Hodgkin and Huxley had postulated that the inactivation

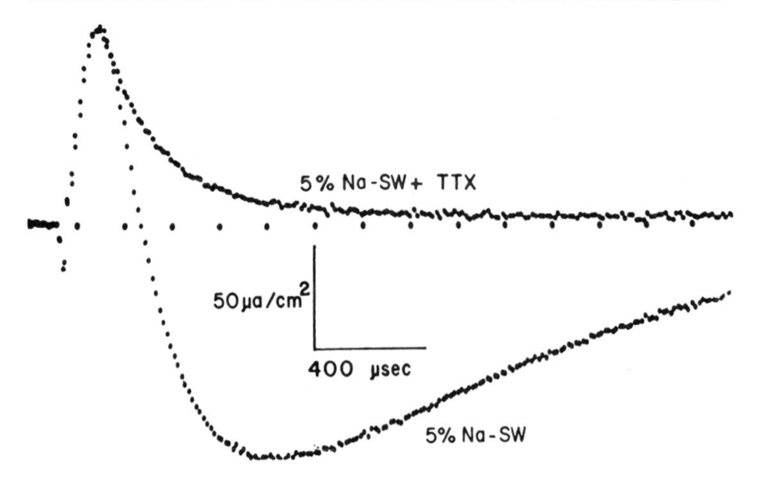

Figure 5. *Gating current and Na channel activation in squid axon membrane. The initial upward (outward) current is gating current, which results from voltage-driven conformational changes in the Na channel protein. The conformational changes lead to opening of Na channels, and an inflow of sodium ions, which obscure the later part of the gating current. Sodium current is much reduced because all but 5% of the external sodium was replaced by an impermeant ion. In the upper trace, sodium current was blocked by the application of tetrodotoxin, a powerful poison from the puffer fish. From Armstrong and Bezanilla, 1974.*

'gate' was, like the activation gate, directly sensitive to voltage. This meant that the closing of the inactivation gate should be associated with a gating current, but we were unable to detect this current. We proposed, instead, that the inactivation gate is not directly affected by voltage, but appears to be voltage dependent because it can close only after activation of the channel. This led to visions of a molecular plug that diffuses into the channel after it opens, a picture that is still generally accepted. The molecular identity of the plug remains to be worked out.

The study of gating current also made it possible to predict some of the salient features of the Na channel molecule (Armstrong, 1981) several years before its sequence was determined by Numa and his coworkers (Noda *et al.*, 1984). The sodium channel is a single large peptide, with approximately 2000 amino acids. Although models based on the sequence are still speculative, it seems likely that a central channel is surrounded by four 'domains,' and in each domain the sequence is similar but not identical (Fig. 6). The similarity of

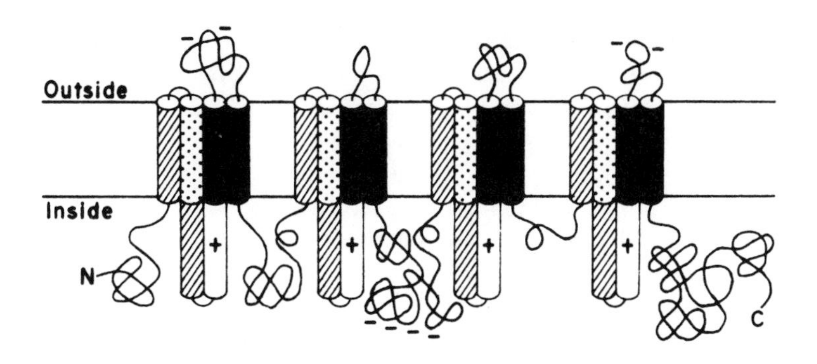

Figure 6. *A schematic indication of the transmembrane crossings of the sodium channel protein. The protein has four similar domains, and in each there are (in this model) four transmembrane crossings by α–helical regions of the peptide (drawn as cylinders). Subsequent models predict six or more crossings. The α–helix marked with a plus has a net positive charge due to repeated arginine or lysine residues, and is thought to lie within the membrane, where it serves, thanks to its charge, as a sensor of membrane voltage. From Noda* et al., *1984.*

sequence within the domains suggests that a primitive ancestral channel may have been made from four identical subunits. Over the course of time, these subunits were stitched together to form a single peptide with four 'domains,' and some differences evolved in the sequence of the domains. In each domain, the peptide winds back and forth across the membrane six to eight times. A striking feature of the sequence is the presence within each domain of periodically spaced, positively charged residues, arginines or lysines, which probably form an α-helix with a net positive charge. This helix is thought to cross the membrane, and because it has a positive charge, it will tend to move when the membrane field is altered. It is this helix that, in all probability, is responsible for the voltage dependence of the sodium channel, and its movement generates gating current. I am pleased to say that both the domain structure of the sodium channel and the existence of a helix with periodic positive charges were predicted from the gating current measurements (Armstrong, 1981) some years before the sequence was determined.

A potassium channel has also been sequenced, from the fruitfly. Interestingly, this peptide is much smaller than the sodium channel peptide, and contains only one repeat of the positively charged arginine-lysine sequence. It is thought that a K channel is formed from the coalescence of four of these peptides around the channel, yielding a structure similar to the sodium channel, but with identical subunits rather than domains. For some reason, the subunits of the K channel have not been stitched together.

The past century thus has seen the understanding of cellular electrical properties progress from a set of astonishingly good guesses regarding cable properties and membrane permeability changes to a clear view of the role of Na and K channels in action potential propagation. In this evolution the work of Cole, Hodgkin, and their coworkers is of first importance. There are still many questions. Eventually one can look forward to detailed x-ray crystallographic information that will help elucidate difficult questions like the origin of selectivity and the details of the conformational changes during channel gating. The assault on these questions is proceeding at a rapid pace, hopefully worthy of the standards set by Cole and Hodgkin.

Literature Cited

Armstrong, C. M. 1981. Sodium channels and gating currents. *Physiol. Rev.* **61**: 644-683.

Armstrong, C. M., and F. Bezanilla. 1973. Currents related to the movement of the gating particle of the sodium channels. *Nature* **242**: 459-461.

Armstrong, C. M., F. Bezanilla, and E. Rojas. 1973. Destruction of sodium conductance inactivation in squid axons perfused with pronase. *J. Gen. Physiol.* **62:** 375-391.

Bernstein, J. 1902. Untersuchungen zur Thermodynamik der bioelektrischen Strome. *Erster Theil. Pflugers Arch.* **82:** 521-562.

Cole, K.S. 1968. *Membranes, Ions, and Impulses*. University of California Press, Berkeley and Los Angeles, CA.

Cole, K. S., and H. J. Curtis. 1939. Electrical impedance of the squid giant axon during activity. *J. Gen. Physiol.* **22:** 649-670.

Cremer, M. 1900. Uber wellen und Psuedowellen. *Zeit. Fur Biol.* **40:** 393-418.

Hermann, L. 1905. Beitrage zur Physiologie und Physik des Nerven. *Arch. ges. Physiol.* **109:** 95-144.

Hille, B. 1984. *The Ionic Channels of Excitable Membranes.* Sinauer, Sunderland, MA.

Hodgkin, A. L. and A. F. Huxley. 1952. A quantitative description of membrane current and its application to conduction and excitation in nerve. *J. Physiol. (London)* **117**: 500-544.

Hodgkin, A. L., A. F. Huxley, and B. Katz. 1952. Measurement of current-voltage relations in the membrane of the giant axon of *Loligo. J. Physiol. (London)* **116**: 424-448.

Hodgkin, A. L., and B. Katz. 1949. The effect of sodium ions on the electrical activity of the giant axon of the squid. *J. Physiol. (London)* **108:** 37-77.

Kelvin, Lord (William Thompson). 1855. On the theory of the electric telegraph. *Proc. Roy. Soc. (London)* B **135**:506-534.

Keynes, R. D., and E. Rojas. 1974. Kinetics and steady-state properties of the charged system controlling sodium conductance in the squid giant axon. *J. Physiol. (London)* **239**: 393-434.

Mitchell, P. 1961. Coupling of phosphorylation to electron and hydrogen transfer by a chemi-osmotic type of mechanism. *Nature* **191**: 144-148.

Mullins, L. J. 1959. An analysis of conductance changes in squid axon. *J. Gen. Physiol.* **42:** 817-829.

Nachmansohn, D. 1966. Chemical control of the permeability cycle during the activity of excitable membranes. *Ann. N.Y. Acad. Sci.* **137:** 877-900.

Narahashi, T., J. W. Moore, and W.R. Scott. 1964. Tetrodotoxin blockage of sodium conductance increase in lobster giant axons. *J. Gen. Physiol.* **47**: 965-974.

Nernst, W. 1888. Zur Kinetik der Losung befindlichen Korper: Theorie der Diffusion. *Z. Phys. Chem.* Pp. 613-637.

Noda, M., S. Shimizu, T. Tanabe, T. Takai, T. Kayano, T. Ikeda, H. Takahashi, H. Nakayama, Y. Kanaoka, N. Minamino, K. Kangawa, H. Matsuo, M. A. Raftery, T. Hirose, S. Inayama, H. Hayashida, T. Miyata, and S. Numa. 1984. Primary structure of electrophorous electricus sodium channel deduced from cDNA sequence. *Nature* **312**: 121-127.

Williams, L. W. 1910. *The Anatomy of the Common Squid,* Loligo pealii. Brill, Leiden, The Netherlands.

Young, J. Z. 1936. Structure of nerve fibres and synapses in some invertebrates. *Cold Spring Harbor Symp. Quant. Biol.* **4:** 1-6.

Selig Hecht
(1892 - 1949)

George Wald
(1906 -   )

Establishing the Molecular Basis of Vision: Hecht and Wald

Introduction by John E. Dowling

Selig Hecht and **George Wald** were giants in the study of photoreception and vision. Both spent many summers working at the MBL, and a number of their fundamental contributions came from research carried out here in Woods Hole. Selig Hecht's studies were of paramount importance in the development of modern concepts of visual mechanisms; they provided the first evidence that visual phenomena can be explained in terms of physics and chemistry. George Wald's research focused on the chemistry of the light-sensitive visual pigments, and he explained at the molecular level many of Hecht's concepts.

Selig Hecht was born in Austria in 1892. He came to this country as a child and grew up in New York City. After obtaining a B.S. degree from the City College of New York in 1913, he entered Harvard for graduate training in biology with G. H. Parker. Parker was another Woods Hole giant, who first came here in the summer of 1888 to work at the Fisheries Laboratory. In 1918, he transferred his research to the MBL and it was at about this time that Hecht, with his newly won doctorate, also began spending summers in Woods Hole. In a series of classic quantitative studies on the physiology of the light responses in the tunicate, *Ciona*, and the clam, *Mya*, carried out between 1917 and 1922, Hecht revealed essential features of the photoreceptor process. This research led to the formulation of a photochemical scheme involving the visual pigments that appeared to explain much of the visual behavior of these simple organisms, including the loss of visual sensitivity in the light (light adaptation) and its recovery in the dark (dark adaptation). The basic idea of this scheme was that photoreceptor sensitivity relates to visual pigment concentration in the photoreceptor cells. In the light, the concentration of photopigment declines to a steady state level, with a concomitant loss of visual sensitivity, whereas in the dark, visual pigment is restored to maximum levels with the recovery of visual sensitivity.

For most of his career, Hecht was Professor of Biophysics at Columbia, until his untimely death in 1948 at the age of 55. During his time at Columbia, Hecht devoted himself primarily to the study of human vision. He applied many of the ideas gained from the study of marine invertebrates to this research. Thus, he eventually explored not only light and dark adaptation, both in man and in *Mya,* but also intensity discrimination, spectral sensitivity, and flickering light resolution.

One of Hecht's students at Columbia was George Wald, who was to set many of Hecht's concepts on the visual pigments into precise chemical terms. Early in his career, Wald showed that the visual pigment molecule is a complex of protein and a slightly altered, or oxidized form of Vitamin A (Vitamin A aldehyde). Beyond this fundamental discovery, which was one of the first instances that a biochemical role for a vitamin had been established, Wald and his colleagues made innumerable contributions to our knowledge of the biochemistry of vision. This work included extensive studies on the chemistry of the rod pigment, rhodopsin, and the extraction and characterization of the first known cone pigment, iodopsin. Wald and his collaborators discovered the role of *cis-trans* isomerization in the visual process, demonstrating for the first time that such molecular transformations play a role in biology. In addition, they provided important insights on the diversity of visual pigments, Vitamin A deficiency, visual adaptation, color vision, and the cone pigments in primates. No one has contributed more to the understanding of the visual pigments and their relation to vision than George Wald. For his enormous contributions Wald received the Nobel Prize in 1967.

Wald was profoundly influenced by Hecht and, when accepting the Proctor Medal in 1955, spoke of him: "Hecht was a great teacher and physiologist. Also he was one of those rare persons who sets a standard both at work and at leisure. I was fortunate in having his instruction and later his friendship. I saw too little of him after leaving his laboratory, but I felt his presence always. What I did or said or wrote was in a sense addressed to him." Wald left Hecht's lab in 1932, and then spent two years as a National Research Council Fellow in Biology. In 1934, he assumed his first academic position as tutor in Biochemical Sciences at Harvard. He remained all of his academic career at Harvard, becoming Professor of Biology in 1940.

Meredithe L. Applebury received a B.S. in Chemistry from the University of Washington and a Ph.D. in Biochemistry/Biophysics from Yale University. Following postdoctoral training in fast kinetics at the Max Planck Institute in Gottingen, West Germany, she joined Princeton University's Department of Biochemical Sciences. There she developed a research program in primary processes of vision and the biochemistry of visual transduction. In later years, she initiated work in the molecular genetics of visual proteins. From Princeton Dr. Applebury moved to Purdue University, where she examined the evolution of visual pigments. She has recently joined the faculty of the University of Chicago as Professor of Ophthalmology and Visual Sciences and of Pharmacological and Physiological Sciences.

Establishing the Molecular Basis of Vision: Hecht and Wald

Meredithe L. Applebury

University of Chicago

MILESTONES IN SCIENCE WARRANT CELEBRATIONS. Such an occasion has arrived following some seventy odd years of research that unraveled one of the great problems of physiology—*the molecular basis of visual signaling*. The achievement rests principally on the fundamental contributions made by two great scientists, Selig Hecht and George Wald. Many of their summers were spent at the Marine Biological Laboratory in Woods Hole, and some of their key experiments were performed here. It was these two who posed the correct questions and who carried out the experiments that offered the insights that led to the unfolding of a remarkable story. It is a privilege to honor these men as part of the centennial celebration of the Marine Biological Laboratory.

The chronicle of events that led to a description of visual signaling during this century can be recounted in three parts: in 1917, Selig Hecht demonstrated that a photosensory response of an organism can be accurately analyzed by the fundamental laws that govern a photochemical reaction. His career was devoted to establishing the qualitative and quantitative basis for investigation of the problem of vision in terms of physics and chemistry. These insights guided all subsequent investigations that ultimately defined the molecular basis of vision.

George Wald was Selig Hecht's student. Wald began a decisive attack on identifying the molecular components of visual processing. His stellar success led to a definitive identification of the substance that absorbs light and undergoes photochemical transformation, the visual pigment rhodopsin. He outlined in detail the molecular process by which photodetection is accomplished. He correctly proposed the mechanism by which this photochemical event could be amplified to create a neural response. In the late 1970s and the early 1980s, the biochemical and physiological efforts of many colleagues in the field completed the description of the visual signaling pathways in vertebrates (Fig. 1).

Both Hecht and Wald drew on a general belief that the fundamental principles of photosensory signaling are common to all organisms that respond to light. Both expended much effort to demonstrate the validity of this concept. What could be functionally characterized in one species was quickly cross documented for other species in other phyla. In the late 1970s, the power of molecular genetics took hold in the field of vision. The tools of recombinant DNA technology were applied to identify the molecules in the visual pathways of eucaryotes as diverse as a single cell or a human being. These studies have shown that the underlying molecules of photosensory signaling are remarkably similar in structure and function from species to species.

Constructing a Molecular Scheme for Photoreception
1917 - 1947 Selig Hecht

The concept that vision could be a photochemical process took form in the 1800s. In 1842, Moser first suggested that light might induce a series of chemical changes in the retina like the processes of photography. Some 35 years later, Boll reported that if the retina is isolated in the dark it has a reddish color. When exposed to light, it slowly bleaches to yellow. Boll named the bleaching substance Sehrot (visual red). This light-induced color change of a substance provided support that vision might be a photochemical process. A year later in 1877, Friedrich Wilhelm Kühne actually extracted a reddish pigment from the retina with a solution of bile salts and showed that it bleached in the light. Kühne named the pigment Sehpurpur (visual purple). He stressed that its reaction to light was likely to initiate the signaling process of vision.

When Selig Hecht began his work, the nature of the chemical processes initiated by light were still obscure. Hecht launched his career in photoreception and vision by choosing the right organism to study. As a student at Harvard, he had worked on the physiology of the ascidian, *Ascidia atra*. Upon finishing, he took a summer fellowship at the Oceanographic Institution in La Jolla and chose another ascidian, *Ciona intestinalis*, for the study of light sensitivity and photomovements. *Ciona* has a simple photosensory response that is free of higher order complexities. The tube-shaped animal branches at one end into two atrial siphons. At the base of the two siphons there

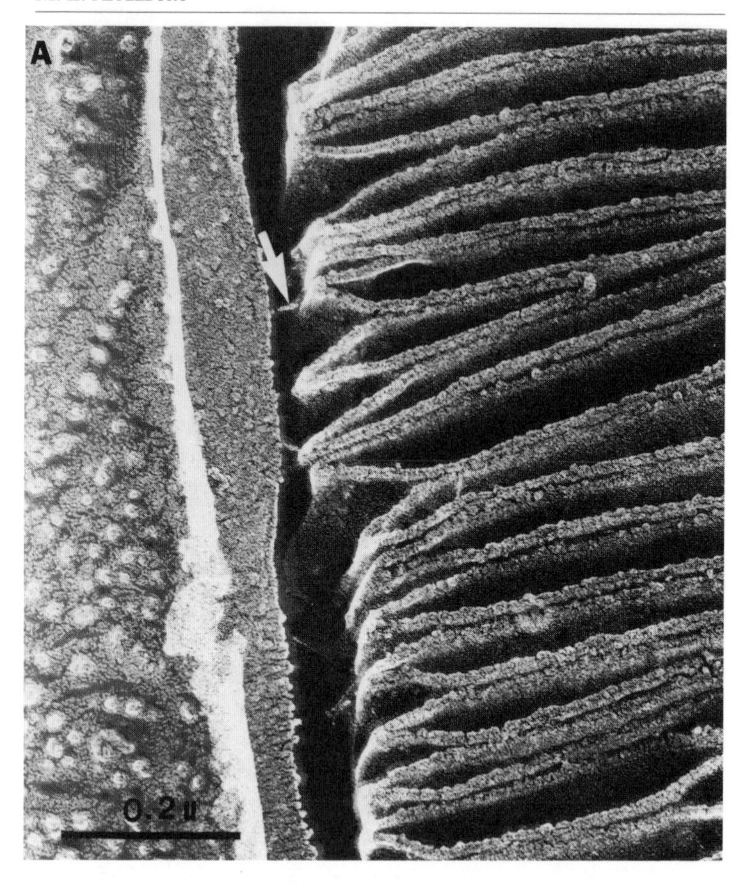

Figure 1. *Visual signaling in vertebrate rod photoreceptor cells. (A) Rod outer segments. A vertebrate rod photoreceptor is a specialized cell for detecting light and triggering a signal that is processed and sent to the brain. The molecules for absorbing light and signaling are housed in the outer segment of the cell. This region consists of stacks of membrane discs, each carrying many copies of the molecular signaling machinery, encased by a plasma membrane (deep-etched freeze fracture rod outer segment by permission of D. Roof).*

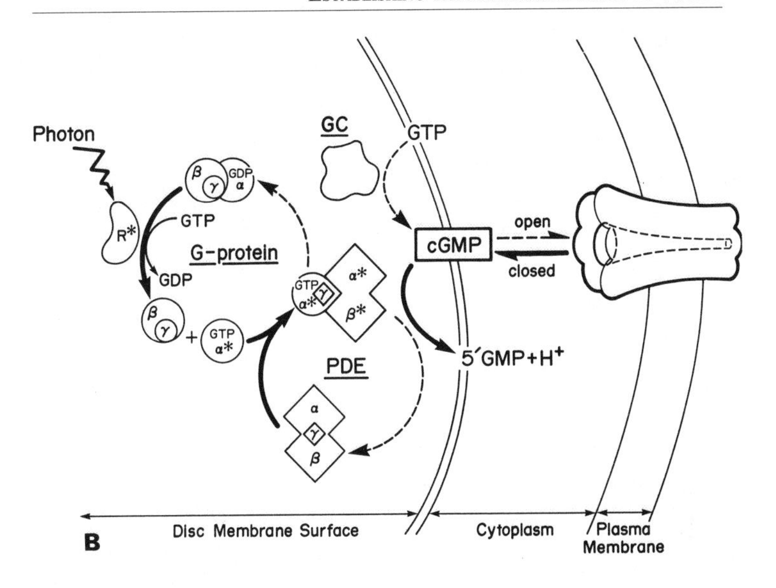

*(B) The molecular basis of visual signaling. A schematic diagram showing the molecular components of visual signaling present in the rod outer segment. Rhodopsin is an integral membrane protein consisting of a protein that covalently binds the chromophore 11-*cis *retinaldehyde, a derivative of Vitamin A. Upon absorbing light, rhodopsin is activated (**R***) and thereby initiates a set of reactions carried out with proteins bound to the membrane disc surface, G-protein and cGMP phosphodiesterase (PDE). **R*** catalyzes the exchange of GTP for GDP bound to the α–subunit of G-protein (also called transducin). In the presence of GTP, the α-subunit dissociates from the βγ–subunit and goes on to activate the PDE by displacing or removing its small inhibitory subunits, γ. The cascade of events is amplified because each **R*** can catalyze the activation of several hundred G proteins and each activated PDE can hydrolize ~ 1000 cGMP per second. Thus, cGMP is rapidly depleted. As the cytoplasmic cGMP level drops, cGMP bound to the ion channel dissociates and the channel closes resulting in membrane hyperpolarization. This change is felt at the synaptic level of the cell and the information is transmitted to the brain. To reset the dark-adapted state, the active rhodopsin is phosphorylated and inhibited through interaction with a 48K protein (not shown), Gα hydrolyzes the GTP and recombines with Gβγ thereby releasing the activated state of PDE. Guanyl cyclase (GC) returns the cGMP to resting concentration.*

is a neuronal mass called the intersiphonal network that is light sensitive. Upon photic stimulation, the siphons vigorously retract. Hecht studied the reaction time, the effects of intensity, and the dark adaptation of this response.

Armed with a stop watch and a meter to measure light intensity in candle meters, Hecht would enter a dark room with his animals about 11:00 in the morning and work until 10:00 in the evening. The response itself (retraction) was relatively constant, requiring about 1.7 s for completion. But the period between photic stimulation and retraction was proportional to the intensity of light—the lower the intensity, the longer the reaction time. From an analysis of his data, Hecht was able to demonstrate that the product of the intensity and the reaction time was a constant (Fig. 2A). *This relationship is a fundamental property of a photochemical reaction.* A corollary is that one can assess the amount of photochemical substance that must be produced to cause a response. The observations were captured in the fundamental law established by Bunsen and Roscoe in 1862, $E=k \cdot I \cdot t$, where E is the photochemical effect, k is a constant that reflects the cross sectional capture of light by the substance and the quantum yield with which it undergoes reaction, I is the intensity of light, and t is time (Hecht, 1918a,b).

Hecht then turned his attention to *adaptation.* Under constant low level light, *Ciona* remains expanded and responds only to relatively bright light. To regain its sensitivity to low level stimulation, *Ciona* must undergo a process of dark adaptation. The process is slow (~2 h), but as the sensitivity improves, the response becomes faster and faster. Hecht intuited that the photochemical substance is slowly reforming to rebuild a pool that when sufficiently large gives rise to a maximum response.

In the fall of 1917, Hecht took a position as Assistant Professor of Biochemistry in the Medical School of Creighton University in Omaha, Nebraska, where he began drafting his publications. Each summer thereafter, Hecht traveled to Woods Hole for research. To confirm the generality of his analysis, he refined his observations of sensory processes in the mollusk, *Mya arenaria*, which retracts its double siphon when illuminated (Hecht, 1919). After fellowship years in European laboratories to further extend his work, Hecht joined Columbia University in 1926. By then he had formulated the basic chemical mechanism for vision that would guide the future work on the underlying molecular components:

$$\mathbf{S} \underset{\text{"dark"}}{\overset{\text{light}}{\rightleftarrows}} \mathbf{P} + \mathbf{A}$$

The photosensitive area of *Ciona* or *Mya* contains a substance **S**. Upon absorption of light, **S** forms two products capable of eliciting the physiological withdrawal response. To account for all properties, particularly dark adaptation, these products are also the precursors **P** and **A** of the photosensitive substance. The system is reversible and forms a sensory equilibrium.

Although today we know the system has many more intermediates, Hecht had formulated the essence of the photosensory process involved in detection and adaptation. He recognized the significance of choosing the right organism for studying a basic phenomena. He thus gained an understanding that enabled him to examine the processes in other organisms. As he stated in 1921, "A careful analysis of the data of dark adaptation in terms of principles discovered in these investigations has shown that dark adaptation and photoreception in the human retina are fundamentally similar in principle to the process in *Mya* and *Ciona*. As a result there has been opened a new field of investigation in retinal photochemistry which may some day enable us to possess a reasonable theory of vision." (Hecht, 1921). Hecht turned his attention to the photophysical properties of human vision to prove just this statement.

To identify the photochemical substance **S** that serves as the photosensitive pigment in the human eye, Hecht and Williams (1922) set about determining an accurate action spectrum for vision. They very carefully calibrated the energy spectrum of their light source. To examine the rod-mediated vision, they used very low levels of light and measured the energy needed to elicit a detectable response in the human eye. By plotting the reciprocal of the threshold energy (which is proportional to the amount of light absorbed that triggers the response) as a function of wavelength, Hecht and Williams obtained an action spectrum for human rod vision (Fig. 2B). With a modest correction, the spectrum matched that of the substance which could be extracted from the retina. The molecular substance that triggered a visual response was the substance called visual purple that Kühne had described.

In 1941-42, Hecht and his coworkers Simon Shlaer and Maurice Pirenne published a second classic experiment that defined the need for amplification in the visual pathway. Hecht was intrigued with determining the minimum number of photons that can impinge upon the retina and still elicit the photoreceptor response. To do this, he and his colleagues first had to accurately determine the level of light that reached the surface of the cornea, calculate what was absorbed by the cornea, and correct for the amount of light that was actually absorbed by the visual cells. The three investigators then analyzed their own responses. This must be done on a statistical basis because each flash of light delivers a small but variable number of photons and each response elicited is stochastic. By analyzing the number of responses as a Poisson distribution, the number of photons actually detected could be determined. Hecht could see, on average, a minimum of six photons, Shlaer seven photons, and Pirenne five (Fig. 2C).

The area of the retina on which the stimulus fell encompassed about 500 photoreceptors. If 6 photons are able to trigger a response in the area of 500 cells, a single rod photoreceptor must be able to capture 1 quantum of light and respond. As Hecht said, *"To see, it is necessary for only one quantum of light to be absorbed by some five to fourteen retinal rods."* Only one visual purple molecule need be stimulated. The remaining role of the rod cell must be devoted to amplifying this photochemical effect to a magnitide sufficient to initiate an electrical signal capable of being transmitted through the retina to the brain. This fundamental property sets the nature and boundary conditions of how visual transduction works. The principle guided the biochemical work done in the 1970s, which described the mechanism of signal amplification.

The Molecular Basis of Phototransduction
1927 - 1977 George Wald

George Wald was a student of Selig Hecht. He studied visual physiology in Hecht's laboratory at Columbia from 1927 to 1932. Wald contributed studies on the visual functions of *Drosophila*, confirming the ever widening knowledge that the basic principles of vision were general throughout nature. Hecht had a profound influence on Wald. As Wald states, "I left Hecht's laboratory with a great desire to lay hands on the molecules for which {**S, P, and A**} were symbols." (Wald, 1968). He left Columbia for Europe in 1932 on what he called his "Wanderjahr," to study and experience research in

A.

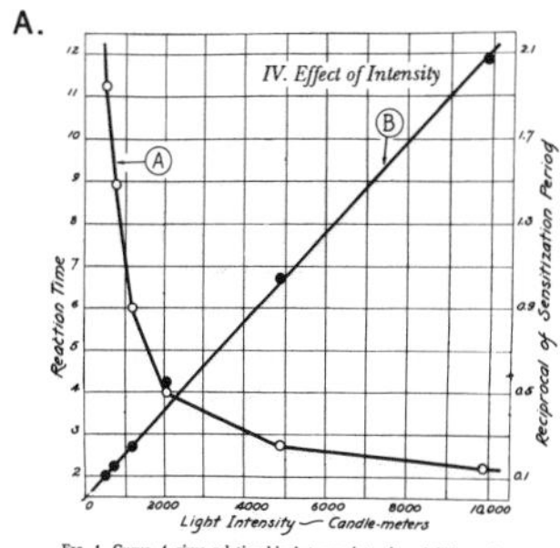

FIG. 1. Curve *A* gives relationship between intensity of light and reaction time of *Ciona*. This curve is an hyperbola as is shown by Curve *B* representing the relation between the reciprocal of the sensitization period and intensity.

B.

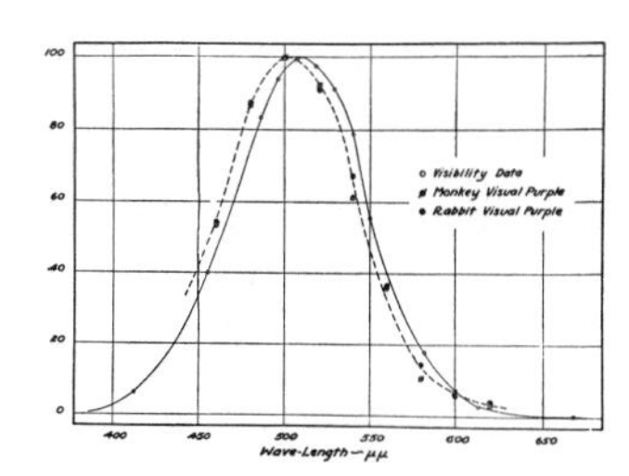

FIG. 7. Relation between absorption spectrum of visual purple in bile salts solution and absorption spectrum of sensitive substance in the rods as given by the visibility curve of Fig. 5. Though the two curves are identical, the visibility curve is shifted 7 or 8 μμ toward the red, as would be expected in terms of Kundt's rule.

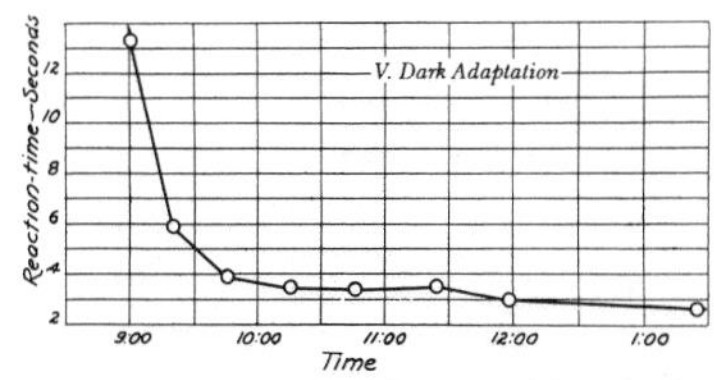

FIG. 2. Course of dark adaptation of *Ciona* represented by reaction time at different periods of sojourn in dark room.

C.

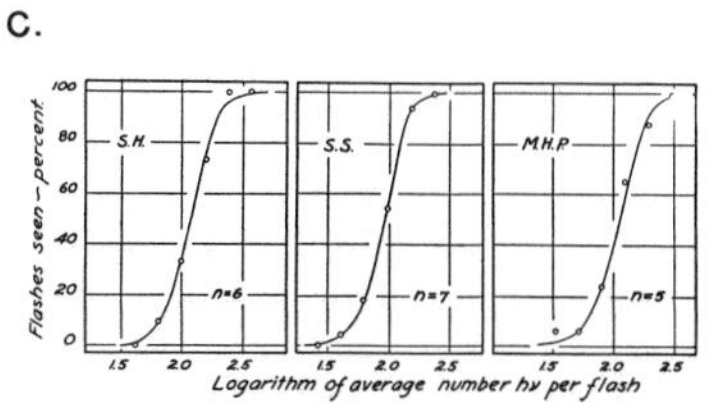

FIG. 7. Relation between the average energy content of a flash of light (in number of $h\nu$) and the frequency with which it is seen by three observers. Each point represents 50 flashes, except for S.H. where the number is 35. The curves are the Poisson distributions of Fig. 6 for *n* values of 5, 6, and 7.

Figure 2. *The photochemical nature of sensory processes. Experimental figures taken from the publications of Selig Hecht. (A) (Fig. 1) The response time of siphon withdrawal in the tunicate* Ciona *was studied as a function of intensity to show the response obeyed the fundamental law of a photochemical reaction, the product of the intensity and time of stimulation is a constant. (Fig. 2) After bright light stimulation, the sensitivity to light is achieved only after dark adaptation. The gradual return of a maximal response time is characteristic of a chemical reaction. Figures taken from Hecht (1918b). (B) (Fig. 7) The action spectrum of the substance that elicits a visual response is nearly identical to the absorption spectrum of the substance called visual purple (rhodopsin) that has been extracted from the monkey or rabbit retina. Subtle corrections needed to be made for absorption of the cornea. Taken from Hecht and Williams (1922). (C) (Fig. 7) Psychophysical studies of the threshold of visual response are analyzed by using a Poisson distribution to fit the number of flash stimuli detected as a function of the number of photons delivered. The classic experiment illustrates that a rod photoreceptor cell is able to detect a single photon and trigger a neural response. Taken from Hecht, Shlaer, and Pirenne (1942). (Reproduced from* J. Gen. Physiol., *1918, Vol. 1, pages 147-166; 1922, Vol. 5, pages 1-33; and 1942, Vol. 25, pages 819-840 by copyright permission of the Rockefeller University Press.)*

the laboratories of three illustrious European Nobel laureates, Warburg, Karrer, and Meyerhof.

In the period of a year, Wald chemically identified the substance **A**. In Warburg's lab in Berlin/Dahlem, he extracted Vitamin A from the retina; he went quickly to Karrer's lab in Zurich, where the structure of Vitamin A had been established, to complete the identification (Wald, 1933). Moving on to Meyerhof's lab, he demonstrated that a derivative of Vitamin A, retinaldehyde (retinene), was liberated from visual purple upon the absorption of light (Wald, 1934). He spent the last phase of this scientific journey at the University of Chicago where he developed these observations into a complete story (Wald, 1934-35).

Wald's discovery of a role for Vitamin A in vision documented one of the first biochemical functions for a vitamin. The work ultimately brought him a Nobel Prize, but by then he had added vastly to our understanding of the molecular basis of visual transduction. He joined Harvard in 1934 and rose through the ranks of academia. He spent his summers in Woods Hole studying vision in all sorts of creatures. He experimentally confirmed the concept of the visual cycle, biochemically characterized visual purple, mapped the photochemical processes that visual purple undergoes upon the absorption of light, identified the cone cell pigments, and confirmed the generality of visual mechanisms at the molecular level by showing that Vitamin A aldehyde is the chromophore in visual pigments of organisms of all types. His contributions were vast. The following highlights only some of the work that turned vision physiology into a discipline of vision biochemistry (Wald, 1968).

Hecht's formula for the photochemical process of vision dictated that there must be a sensory equilibrium between substance **S** and product **P**. To establish this "visual cycle," Wald needed to show that Vitamin A was not only the product of photoreaction, but the precursor of the photoreactive substance **S**. Tracking the presence of Vitamin A in the eye, Wald found it easy to extract the compound from the pigment epithelium (*i.e.,* the cells that nurture the retina and provide a barrier between the arterial blood supply and the photoreceptor cells of the outer retina). But the Vitamin A could not be extracted from the retina itself—unless the retina was first bleached with light. Immediately on exposure to strong light the retina bleaches from red to yellow, and one could then extract a yellow analog of Vitamin A, the oxidized product called retinaldehyde (Fig. 3). If

A.

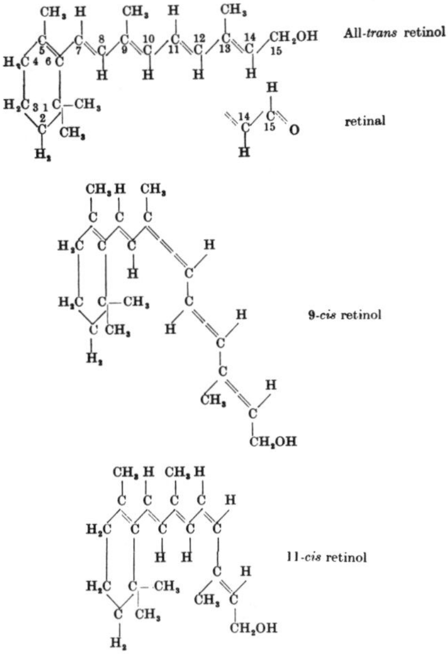

All-*trans* retinol

retinal

9-*cis* retinol

11-*cis* retinol

Fig. 2. Structures of the all-*trans*, 9-*cis* and 11-*cis* isomers of retinol and retinal.

B. CAROTENOIDS AND THE VISUAL CYCLE

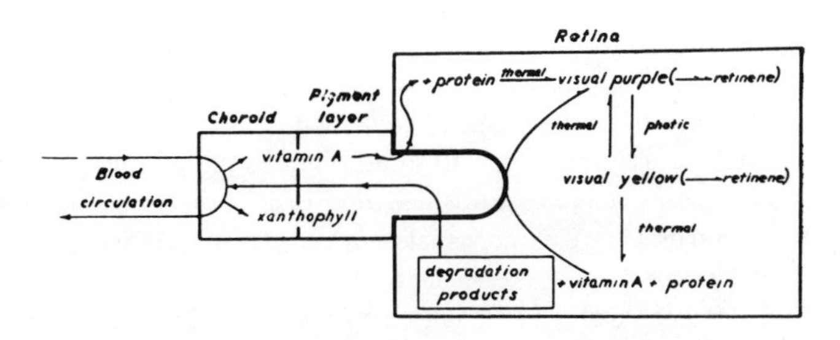

Figure 3. *The role of Vitamin A in vision. (A) The structure of Vitamin A (all-*trans *retinol) and retinaldehyde (retinal). Taken from Wald (1968). (B) The visual cycle. Taken from Wald (1934). (Reprinted by permission from* Nature, *Vol. 134, p. 65, and Vol. 219, pages 800-807. Copyright 1934 and 1968 Macmillan Magazines Limited.)*

isolated retinas were allowed to remain in light or darkness, they bleach to colorlessness as the retinaldehyde is reduced to Vitamin A, the retinol. Allowing the bleached retinas to dark adapt in an eyecup or animal, the visual purple would regenerate and again be capable of undergoing photochemical reaction. Thus examining the isolated retina and its underlying pigment epithelium, Wald described the molecular basis of the visual cycle. When visual purple was bleached, it released the retinaldehyde chromophore. The latter was slowly converted to the retinal alcohol, Vitamin A. During dark adaptation, the retinaldehyde derived from Vitamin A would recombine with a colloidal substance to reform visual purple. The pigment epithelium served as a store of Vitamin A (Fig. 3) (Wald, 1935-36).

Wald spent the next years describing the nature of the colloidal substance that combined with retinaldehyde to form visual purple. The substance had all the properties of a protein; it could be salted out of solution, it was denatured by heat, acids, alkalies, organics, etc. Visual purple was indeed a protein that contained a tightly bound prosthetic group that served as a chromophore. If visual purple were to be a general biological pigment present in all photosensitive organisms, the protein, its bound retinaldehyde, and the general properties of the visual cycle should be present in other species. In each organism Wald examined, many at Woods Hole, he confirmed this generality (Wald, 1935-36; 1936-37; and 1937). One surprise and variation was noted with the freshwater fishes. The properties of their visual pigments were similar, but the absorption spectra were red shifted. The chromophore was Vitamin A-like and called Vitamin A_2 because it turned out to have an extra double bond in the structure that shifted its absorption. Wald named these pigments porphyropsin and designated rhodopsin for the rod pigments to distinguish these two types of "visual purple." Still, the general theme was universal. Each type of retina that was examined contained a visual pigment with a characteristic absorption spectrum ascribed to the presence of the extractable chromophore retinaldehyde. Each retina contained the product and precursor of the visual pigment, Vitamin A, and demonstrated the properties of the visual cycle.

Wald then turned his attention to the photochemical process itself. Following the absorption of light, rhodopsin undergoes a series of thermal reactions called bleaching that results in a colorless photoproduct (Wald, 1938). The products of thermal decay could be trapped by lowering the temperature or shifting the pH. By the late

1940s, Wald was joined by Ruth Hubbard and Paul Brown, two colleagues who were to work closely with him to define the photochemical basis of transduction. They developed a *de novo* regeneration system that enabled them to prove that rhodopsin is synthesized from the protein opsin and a specific isomer of retinaldehyde (Wald and Brown, 1950). By trapping the intermediates and extracting the bound retinaldehyde, they were able to show that the native dark-adapted rhodopsin bound the 11-*cis* isomer of retinaldehyde. Upon bleaching, the chromophore was found in the all-*trans* configuration (Hubbard and Wald, 1952-53). During this period, Wald and his colleagues characterized the visual pigments from cone cells. These pigments were synthesized from opsins of a slightly different character and 11-*cis* retinaldehyde. Their photobleaching processes were analogous to those of rhodopsin (Wald *et al.*, 1954-55). In both types of pigment, light produced an isomerization of the chromophore about the 11-12 double bond. The fundamental step in the transduction process of all visual pigments was solved—*the action of light on this photosensitive material is to isomerize the 11-*cis *retinal chromophore to the all-*trans *state.*

If the chromophore changed its configuration upon absorption of light, the protein must be driven to change and accommodate this new configuration. Wald and his co-investigators began to map out the intermediates of bleaching more carefully, recording the spectrum of each intermediate and the kinetics leading to the dynamic process of thermal decay. At very low temperatures, one could trap a very early intermediate pre-lumirhodopsin (bathorhodopsin); at higher temperatures pre-lumirhodopsin decayed to lumirhodopsin; lumi- decayed to Meta I-, and Meta I- was found in an equilibrium with Meta II-rhodopsin (Fig. 4) (Yoshizawa and Wald, 1963; Wald, 1968). Each intermediate had a characteristic spectrum; each state reflected a slightly altered interaction of the protein with its chromophore. This beautiful system was a rare biological example in which a photochemically induced reaction could be "watched." The scheme for bleaching that was inferred indicated how the absorption of energy could be used to drive a macromolecule into a new state, change its conformation, and thereby activate it for visual signaling. Not only had Wald identified Hecht's **S**, **P**, and **A**, he had mapped the chemical changes underlying visual transduction.

Despite this progress, a fundamental question lay unanswered. How did this photochemical process induce the physiological response, the train of neural events that leads to seeing? What did photoactivated rhodopsin initiate? Governed by Hecht's demonstration that the human eye can detect a single photon, somehow a single rhodopsin molecule must be able to evoke a neural response. Clearly, a large amplification was required. Though the answers were to come in the late 1970s and 1980s, Wald, with marvelous insight, proposed that light-activated rhodopsin might act as a catalytic center. Thus, it could serve as a first step in a series of reactions that would lead to amplification. He made an analogy to the cascade of events that occurred in blood clotting. He closed his publication with the statement "...such a biochemical arrangement for multistage amplification, capable of drawing a large return very rapidly from a minimal input...seems so applicable to the general problem of excitation...that it would seem worthwhile to look for it {in receptors}." (Wald, 1965).

Indeed, research in the field of vision during the 1970s proved Wald correct. With the detection of cyclic nucleotides in photoreceptors, the refined biochemical techniques for characterizing proteins, and the application of patch clamping to rod cells, visual signaling was solved. The last piece of the story came in 1984, when Fesenko and his colleagues demonstrated that cGMP directly affects the opening of ion channels in the photoreceptor plasma membrane. Rhodopsin, activated by light, triggers a cascade of events leading to the hydrolysis of the second messenger, cGMP (Fig. 1) (for review see Stryer, 1986). As the cyclic nucleotide concentration drops, the ion channels close, leading to hyperpolarization of the plasma membrane. This signal is transmitted via many synaptic relays to the brain—a train of neural events that constitutes "seeing."

Visual Pigments: Part of a Universal Plan for Signaling 1980 - 1988

Hecht and Wald accumulated a wealth of observations to support the dogma that throughout the animal kingdom, all visual systems represent variations on a central theme. Hecht had started his work with *Ciona* and had drawn on the insights he gained to study human vision. Wald had repeatedly shown that the photochemical process of vision was functionally similar in mammals, amphibians, fish, birds, annelids, and insects. A strong understanding of the structure and dynamics of the photochemically induced changes of rhodopsin isolated from bovine rod outer segments was achieved. But retinas from other organisms, from human, *Drosophila*, etc., were either

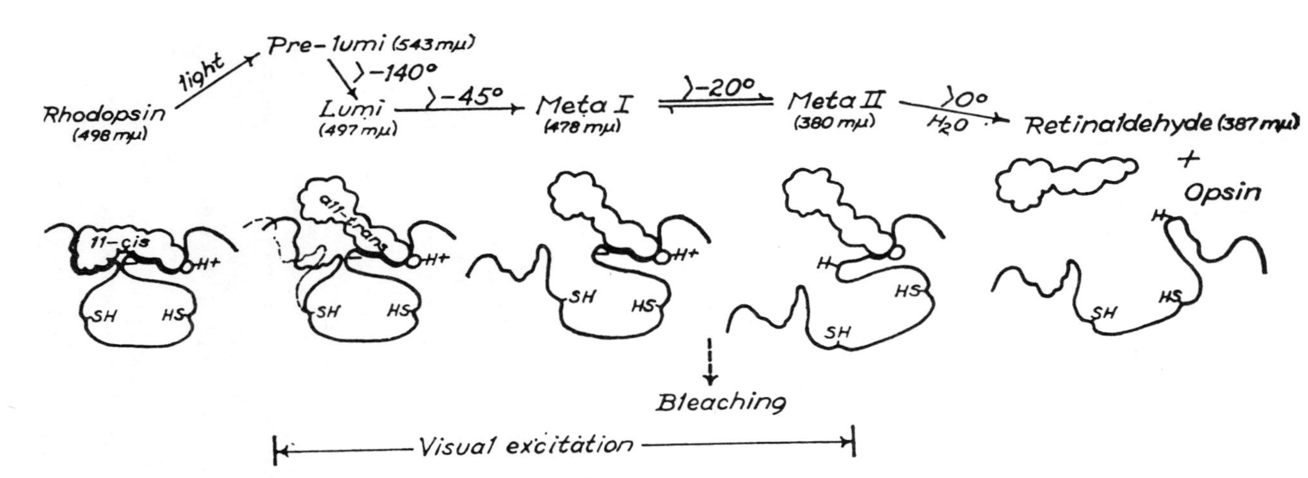

Fig. 6. Stages in the bleaching of rhodopsin. The chromophore of rhodopsin, 11-*cis* retinal, fits closely a section of the opsin structure. The only action of light is to isomerize retinal from the 11-*cis* to the all-*trans* configuration (prelumirhodopsin). Then the structure of opsin opens progressively (lumi and the metarhodopsins), ending in the hydrolysis of retinal from opsin. Bleaching occurs in going from metarhodopsin I to II; and visual excitation must have occurred by this stage. The opening of opsin exposes new chemical groups, including two —SH groups and one H^+-binding group. The absorption maxima shown are for prelumirhodopsin at −190° C, lumirhodopsin at −65° C, and the other pigments at room temperature.

Figure 4. *The molecular process of phototransduction. The intermediates of bleaching. Reprinted by permission from* Nature, *Vol. 219, pages 800-807. Copyright 1968 Macmillan Magazines Limited.)*

inaccessible or too small to yield the necessary amount of material for molecular studies. Moreover, the primary amino acid structure of this membrane-embedded protein was exceptionally difficult to determine and a complete sequence was still unavailable in 1980.

An alternative to the study of visual pigments in all organisms became available with advances in molecular genetics and recombinant DNA technology made in the late 1970s. By 1983, bovine rhodopsin was cloned and sequenced in our laboratory, Gobind Khorana's lab at MIT, and by Jeremy Nathans at Stanford. The data provided a confirmation of the bovine protein sequence that had just been published from Ovchinnicov's laboratory in Moscow and by Paul Hargrave in Illinois. Structural detail of one visual pigment was now available for speculating how the chromophore interacted with the protein and what changes in structure might trigger the cascade of events for neural signaling.

The cloned bovine opsin opened an additional possibility. If what Hecht and Wald had proclaimed was true, the visual pigments in all organisms should have basic molecular properties in common. The bovine cDNA clone encoding opsin should cross hybridize with opsins in the genomes from other organisms. This notion proved valid beyond our expectations. Rick Martin used the bovine cDNA as a probe to show that it could detect opsins in the genomes of every organism studied that was known to have a photosensitive response, from human to a blue-green algae, *Chlamydomonas* (Martin *et al.*, 1986). The bovine probe was used by our lab to identify and characterize opsins from mouse, canine, *Drosophila*, and *Chlamydomonas,* in Jeremy Nathan's lab to identify human rod opsin, and the cone red-, green-, and blue-sensitive pigments, and in Tokunaga's lab in Japan to identify the chicken rod opsin. Additional opsins have now been identified from the R7 rhabdomeric photoreceptor and the ocelli of *Drosophila* and from octopus. The visual pigments form a family of signal receptors with remarkable molecular similarity. The genes that encode them are frequently similar in their intron and exon structure and have regions of identity in their nucleotide sequences (O'Tousa *et al.,* 1985; Falk and Applebury, 1987). The proteins are similar in their primary amino acid sequences and predicted folding patterns; regions of the cytoplasmic surface that serves as an interface for the GTP-binding protein (the second protein to be activated in the chain of events) are strikingly similar (Applebury and Hargrave, 1986). Indeed, *all visual pigments are built on a common plan.*

What was perhaps unexpected was that the visual pigments are only a subgroup of an even larger family of proteins. In 1985, Lefkowitz and his colleagues used recombinant DNA technology to clone the β-adrenergic receptor from hamsters. The mechanisms of action of visual pigments and adrenergic receptors have several functional similarities. Both are integral membrane proteins, both are activated by a ligand bound to the protein (rhodopsin may be considered to have poised ligand that is converted to the active form by light), both catalyze the activation of GTP-binding proteins, and both communicate information external to the cell into an intracellular response. When the β-adrenergic receptor was cloned and its primary amino acid sequence deduced from the nucleic acid sequence, the two receptors turned out to be amazingly similar.

In rapid succession, receptors were cloned which encoded proteins that were activated by acetylcholine (muscarinic), the peptides substance **K** and angiotensin, seratonin, and dopamine. In yeast, receptors for mating factors were found to belong to this family. In the primitive single-cell slime mold *Dictyostelium*, a receptor that was activated by cAMP was identified as a member. We can expect to see similar structures for receptors binding prostaglandins, metenkephalins, opioids, substance **P**, histamine, etc. Mammals may well have over 500 different members and homologous counterparts will be found in other species. The visual pigments are but a subclass of the superfamily of signal receptors.

All of these receptors have a basic motif of seven transmembrane segments alternately connected by loops on the extracellular and cytoplasmic side of the membrane bilayer. The transmembrane segments fold in three dimensions to form a ligand binding pocket; the amino acids that line the pocket vary from receptor to receptor to specify the binding of a particular type of ligand, although the general folding pattern is maintained. The loops on the cytoplasmic surface form binding sites for the GTP-binding proteins. This catalytic surface is exposed when the receptor is activated by ligand-binding or light. Loops between transmembrane 1 and 2, 3 and 4, and the region extending from the 7th transmembrane towards the C-terminal are quite similar among the entire superfamily (Fig. 5).

Members of the family have now been found in nearly every phylum, with one exception, the higher plants (Fig. 6). Even here, it is possible that such receptors exist but have not yet been identified. *The similarities in receptor structure across the phylogenetic expanse*

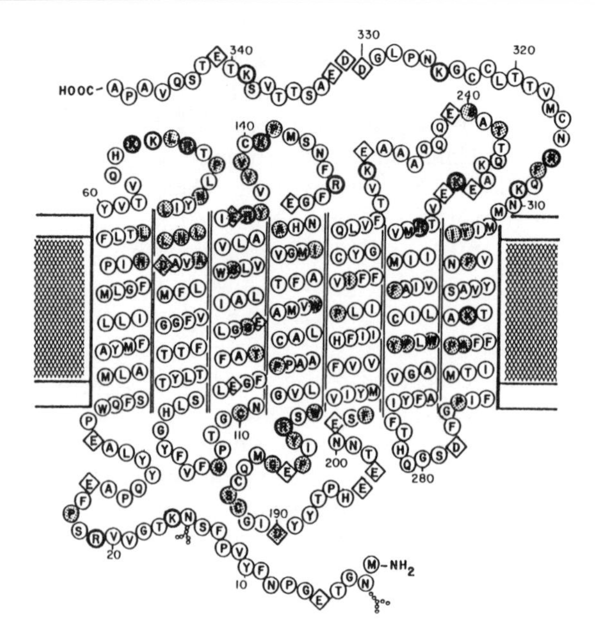
HOOC-
330
340
320
240
140
60
310
110
200
190
280
20
10
M-NH2
bovine opsin

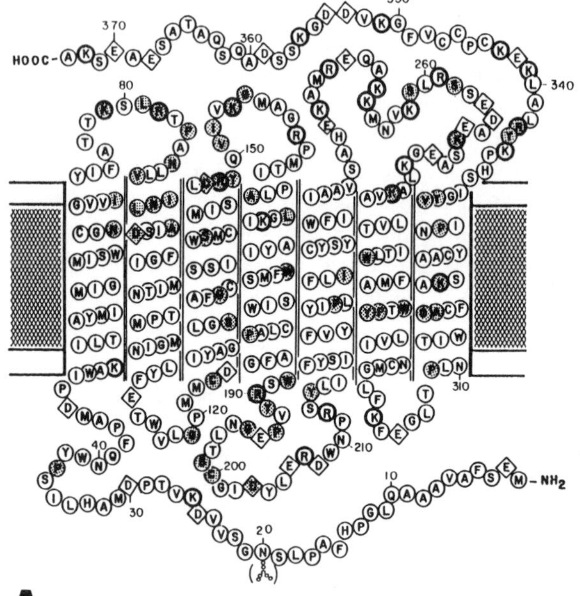
350
370
360
HOOC-
260
80
340
150
190
120
310
210
40
200
10
M-NH2
30
20
A
Drosophila opsin

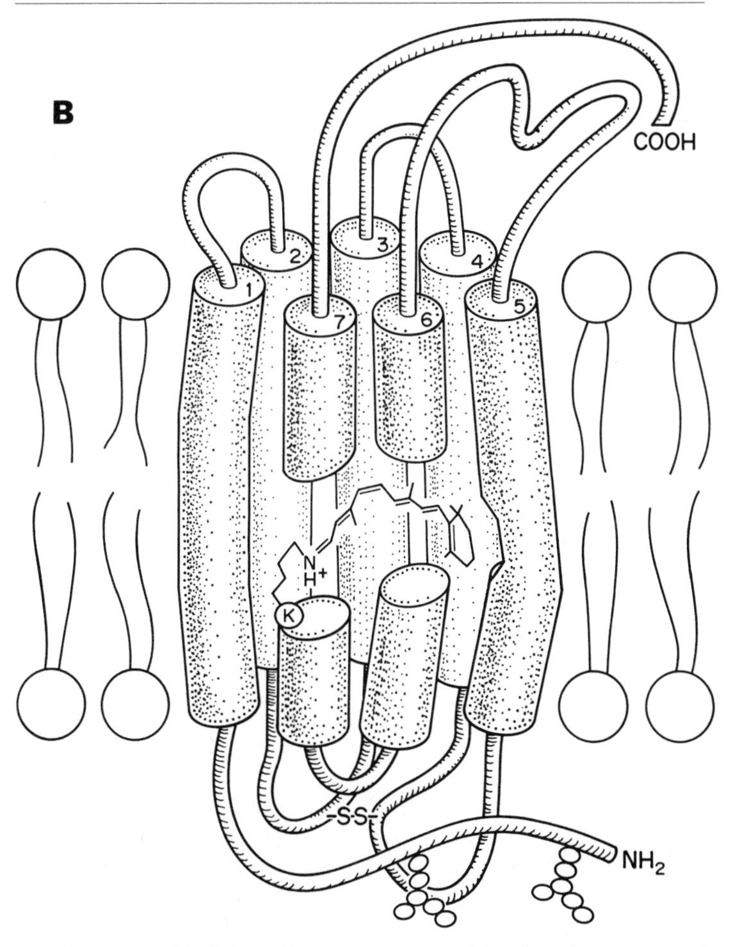

Figure 5. *(A) The primary structure of bovine rhodopsin and* Drosophila *rhodopsin. (B) A model for the tertiary structure of rhodopsin and other signal receptors.*

of eucaryotes imply that the entire family is derived from a common ancestor. The gene was replicated and allowed to drift in its amino acid composition to change its specificity of ligand binding, but not its general mechanism of action. Some 1 to 2 billion years ago, Nature found a remarkably facile structure that could be altered to detect hundreds of different ligands and impart instructions for physiological responses within the cell.

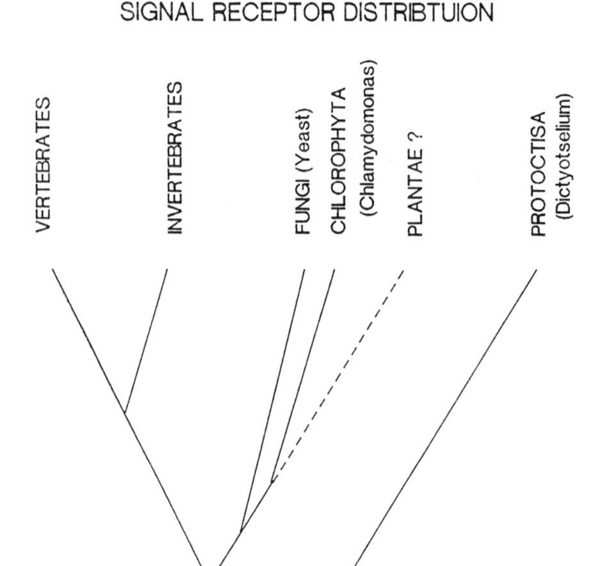

Figure 6. *Presence of visual pigments and signal receptors among the eucaryotes. Both fungi and blue-green algae have signal receptors. Their presence in plants is unknown. The tree indicates evolutionary relationships of the eucaryotes based on sequences of ribosomal RNA.*

Of course, the prototype for this family is the visual pigment. It all started in 1917 with the insight of one young man, who already knew that the tunicates were a good model for the study of photosensory response. And he imparted the fundamental paradigms that enabled his student to identify the molecular nature of the visual pigments. These two never lost sight of the generality of photosensory action throughout all species. Would they, or would they not, have been astounded to know that their fundamental insights addressed one of Nature's universal plans for signal communication?

Ending Notes

These comments were delivered as a Friday Evening Lecture at Woods Hole in June of 1988. To bring the lecture to a close, I chose to tell an anecdote that had been delivered to me by a fellow colleague. Our work in the past few years had left a remarkable impression that visual pigments, and other signal receptors, were indeed built on a common plan. Such molecules could be identified in every eucaryotic organism that had a light response. The family of molecules also extended to other similar signal receptors which detected ligands such as adrenaline, histamine, neuropeptides, and in a very primitive slime mold, cAMP. Certainly, these receptors had derived from some common ancestral gene. Two questions were outstanding. Which came first in evolution, visual pigments or ligand-detecting signal receptors? What would prove true in the last phylum where these molecules have not yet been identified, that is have plants adopted a visual pigment for signaling or have they found other mechanisms more useful?

Aside from the large questions, during much of our work cloning and sequencing visual pigments from animals of all walks of life, I was often bogged in the details. My colleague, Yorum Solomon, a jolly, talmudic scholar from the Weizman Institute, reminded me that I was not seeing the forest for the trees. The answer to the larger questions was there, he said. Indeed in reading sequences of proteins, I simply forgot to punctuate the basic sequence structurally. One of my sequences read:

10 20 30 40 50 60
INTHEBEGININGGODCREATEDTHEHEAVENANDTHEEARTHNOWTHEEARTHWASWIT

70 80 90 100 110 120
HOUTFORMANDVOIDANDDARKNESSWASUPONTHEFACEOFTHEDEEPANDTHESPIRI

130 140 150 160 170 180
TOFGODHOVEREDOVERTHEFACEOFTHEWATERSANDGODSAIDLETTHEREBELIGHT

190 200 210 220 230 240
ANDTHEREWASLIGHTANDGODSAWTHELIGHTANDTHATITWASGOODANDGODIVID

250 260 270 280 290 300
EDTHELIGHTFROMTHEDARKNESSANDGODCALLEDTHELIGHTDAYANDTHEDARKNES

310 320 330 340 350
SHECALLEDNIGHTANDTHEEVENINGANDTHEMORNINGWERETHEFIRSTDAY*

Yorum reminded me that on the first day God said, "Let there be light." And there was light and God saw that the light was good. The photoreceptor was there before the beginning of life.

Following the lecture, Professor Dowling asked Professor Wald if he would like to comment on the history or marvelous unfolding of the molecular basis of vision. Wald stood with a pondering smile. He faced me, then the audience, and gently said:

"On the first day of the Creation God said, 'Let there be light'; and there was light. And God saw that the light was good. And God divided the light from the darkness. And the light he called day, and the darkness he called night. And the evening and the morning were day one. On the third day God created the green plants. Fortunately for the plants, on the fourth day he finally created the sun and the moon.

In the little Jewish villages, the so-called 'shtetls' in Czarist Russia, each village was very proud of the wisdom of its rabbi. In one such village, they came to ask the rabbi, 'Which is more important, the sun or the moon?' 'The moon of course' answered the wise rabbi, 'The sun only shines by day, when it's light anyhow; but what would we do at night without the moon?' The audience was much amused by the sheer ridiculousness of this reply; but the rabbi was quite correct in following the account in Genesis. For God having called for the daylight on the first day of the Creation, when on the fourth day he made the sun and moon, the sun was quite unnecessary since it was already light anyhow; but the moon — ah, yes! — the moon . . . "

Certainly, the plant phyla had homologues of visual receptors ready to detect the light of Creation by the third day; only on the fourth day would the sun arrive to sustain their growth.

Literature Cited

Applebury, M. L., and P. A. Hargrave. 1986. Molecular biology of the visual pigments. *Vision Res.* **26:** 1881-1895.

Boll, F. 1876. On the anatomy and physiology of the retina. *Monatsberichte der Berliner Akademie* **23** (12 November): 783-788.

Boll, F. 1877. On the anatomy and physiology of the retina. *Archiv. f. Anat. and Physiol.* Abt., 4-35. [In translation "Visual Photochemistry. The Beginnings" (eds. T. Shipley and F. Crescitelli) Pergamon, Oxford. Pp. 1247-1265].

Falk, J., and M. L. Applebury. 1987. Molecular genetics of photoreceptor cells. In *Progress in Retinal Research* , N. N. Osborne and G. J. Chader, eds. Pergamon Press, Oxford. **7:** 89-112

Hecht, S. 1918a. Adaptation in photosensitivity of *Ciona intestinalis. Science* **48:** 198-201.

Hecht, S. 1918b. The photic sensitivity of *Ciona intestinalis. J. Gen. Physiol.* **1:** 147-166.

Hecht, S. 1919. Sensory equilibrium and dark adaptation in *Mya arenaria. J. Gen. Physiol.* **1:** 548-558.

Hecht, S. 1921. The photochemistry of the sensitivity of animals to light. *Science* **53:** 347-351.

Hecht, S., and R. E. William. 1922. The visibility of monochromatic radiation and the absorption spectrum of visual purple. *J. Gen. Physiol.* **5:** 1-33.

Hecht, S., S. Shlaer, and M. H. Pirenne. 1942. Energy, quanta, and vision. *J. Gen. Physiol.* **25:** 819-840.

Hubbard, R., and G. Wald. 1952-53. Cis-*trans* isomers of Vitamin A and retinene in the rhodopsin system. *J. Gen. Physiol.* **36:** 269-315.

Kühne, W. 1879. Chemical processes in the retina. [In translation: Visual Photochemistry. The Beginnings, T. Shipley and F. Crescitelli, eds. Pergamon, Oxford, pp. 1273-1316.]. Pp. 235-342 in *Handbuch der Physiologie,* L. Hermann, ed. Vogel, Leipzig, Germany.

Martin, R. L., C. Wood, W. Baehr, and M. L. Applebury. 1986. Visual pigment homologies revealed by DNA hybridization. *Science* **232:** 1266-1269.

Moser, L. 1842. On the process of vision and the effect of light on all substances. *Ann. Phys. Chem., Leipzig* **56:** 177-234.

O'Tousa, J. E., W. Baehr, R. L. Martin, J. Hirsh, W. L. Pak, and M. L. Applebury. 1985. The *Drosophila ninaE* gene encodes an opsin. *Cell* **40:** 839-850.

Stryer, L. 1986. Cyclic cGMP cascades of vision. *Ann. Rev. Neurosci.* **9:** 87-119.

Wald, G. 1933. Vitamin A in the retina. *Nature* **132:** 316-317.

Wald, G. 1934. Carotenoids and the Vitamin A cycle in vision. *Nature* **134:** 65.

Wald, G. 1934-35. Vitamin A in eye tissues. *J. Gen. Physiol.* **18:** 905-915.

Wald, G. 1935-36. Carotenoids and the visual cycle. *J. Gen. Physiol.* **19:** 351-371.

Wald, G. 1935-36. Pigments of the retina. I. The bull frog. *J. Gen. Physiol.* **19:** 781-795.

Wald, G. 1936-37. Pigments of the retina. II. Sea robin, sea bass, and scup. *J. Gen. Physiol.* **20:** 45-56.

Wald, G. 1937. Visual purple system in fresh-water fishes. *Nature* **139:** 1017-1018.

Wald, G. 1938. On rhodopsin in solution. *J. Gen. Physiol.* **21:** 795-832.

Wald, G. 1965. Visual excitation and blood clotting. *Science* **150:** 1028-1030.

Wald, G. 1968. The molecular basis of visual excitation. *Nature* **219:** 800-807.

Wald, G., and P. K. Brown, 1950. The synthesis of rhodopsin from retinene$_1$. *Proc. Natl. Acad. Sci. USA* **36:** 84-92.

Wald, G., P. K. Brown, and P. H. Smith. 1954-55. Iodopsin. *J. Gen. Physiol.* **38:** 623-681.

Yoshizawa, T., and G. Wald. 1963. Pre-lumirhodopsin and the bleaching of visual pigments. *Nature* **197:** 1279-1286.

Haldan K. Hartline
(1903 - 1983)

Stephen W. Kuffler
(1913 - 1980)

Neural Mechanisms of Visual Perception: The Legacy of Hartline and Kuffler

Introduction by Robert B. Barlow

We honor two pioneers in research on the nervous system, **H. Keffer Hartline** and **Stephen W. Kuffler**. Both spent many summers in Woods Hole carrying out studies that gained them wide recognition. For them the Marine Biological Laboratory was a very special place. They spoke often of its invigorating intellectual environment and excellent facilities for investigating marine organisms. Hartline advised students and young colleagues to spend at least one summer in Woods Hole. Of course, he knew full well how difficult it is to spend just one summer there.

Hartline began his research career in the late 1920's about ten years before Kuffler. Both made significant scientific contributions before World War II, but it was during the years following the war that they emerged as leaders in the rapidly growing field of neurobiology. This was indeed a remarkable period. Techniques were being developed to study the physiological properties of single nerve cells by penetrating them with glass microelectrodes and recording their responses with specially designed amplifiers. Progress was rapid, but techniques were far from perfect and success required much patience and a high level of manual skill.

Hartline and Kuffler had both. Anyone who tries to carry out the elegant single fiber nerve-muscle preparation that Kuffler developed in 1940 or attempts to tease apart single optic nerve fibers in the frog retina as Hartline did in 1938 will quickly realize that these men were truly artists. With these techniques, Kuffler demonstrated that acetylcholine sensitivity was localized in the endplate region of the muscle, settling a long controversy, and Hartline showed that cells in the retina possessed a receptive field, that is they gathered visual information over a small region of visual space. Both discoveries had far-reaching consequences. Anyone attempting to repeat these experiments ought to know, for example, that Hartline made his own scissors for dissecting the tiny optic nerve fibers. He invented a tweezer-type microscissors with tips that met exactly when viewed under a microscope. He would test them by cutting through just half of a single cotton fiber, a feat not easily accomplished with the finest surgical scissors available today.

As you may well surmise, Hartline and Kuffler were consummate scientists. They were happiest in the laboratory. It was their home. Although both chaired departments during their careers they preferred to spend time working in the laboratory, leaving as much administration as possible to others. Their enjoyment of life was infectious and set the tone of their laboratories, which in addition to being light, was often chaotic, giving the impression that the course of research was not following any particular plan. However their pursuit of important questions in neurobiology was deliberate and in a quiet, unassuming way they inspired and encouraged all around them, co-workers and students alike, gently guiding their research without appearing to do so.

Our honorees had much in common including, I might add, a marvelous sense of humor. Kuffler often drove his colleagues to despair with an unending string of puns. They tried to ration his puns to one a day without success. Although Hartline was not a punster, he always saw the light side of most everything, including his own research. As many of you may recall, when discussing his research on *Limulus*, Hartline often joked he was studying vision in a blind animal. Levity was abundant in their laboratories.

The scientific backgrounds of Hartline and Kuffler were as different as their personalities were similar (see Barlow, 1986; McMahan, 1990; and Ratliff, 1990). Hartline grew up in Pennsylvania, and learned biology from his father, whom he considered his first and best teacher. Kuffler on the other hand was born in Hungary, didn't much like school, and had no exposure to science in his early years. Hartline entered Lafayette College in 1920 and quickly drew the attention of Professor Beverly Kunkel who, to get Hartline involved in lab work, advised him to go into the woods and collect small pill bugs from the undersides of logs. After several weeks of observation, Hartline reported to Kunkel that he was not certain what could be learned from the little isopods. Kunkel laughed and suggested he return to the laboratory. Hartline did and soon noticed the bugs tended to avoid light. With the meticulous care that was to typify his later work, he studied their visually guided behavior and concluded that "these experiments seem to show a relationship between certain photochemical laws and the phototropism of animals." This study instilled in him a keen interest in the neural events leading from photochemical changes in the eye to changes in an animal's behavior. It set the stage for his career in vision research.

Hartline in this study drew heavily on the photochemical work of Selig Hecht and the animal behavior work of Jacques Loeb, both of whom are also honored in this volume. Hartline had the opportunity to meet Hecht and Loeb at the MBL when, following Kunkel's advice to go to Woods Hole, he enrolled in the Physiology course. They were particularly impressed with his study on pill bugs and after some discussion, Loeb suggested he attend medical school if he wished to pursue basic research. Brimming with enthusiam, Hartline entered Johns Hopkins Medical School in 1923. Much to his dismay he found the medical courses very time consuming, but finally worked out a schedule for lecture work during the day and laboratory work at night. He became fascinated with the evoked retinal potential, the electroretinogram he was able to record from frog, cat, and human, but in time grew weary of the vertebrate eye's complexity and returned to the MBL in 1926 in search of a simpler visual system. It was then he noticed the large jewel-like eye of the horseshoe crab; the rest is history.

The outpouring of results from Hartline's experiments on the horseshoe crab eye was enormous. They touched on almost every aspect of vision and led to the formulation of basic mechanisms of retinal function applicable to many species. Hartline's approach to studying retinal function was reductionistic. Inspired by the pioneering work of Loeb, Hartline tried to interpret biological mechanisms in the simplest terms—according to the laws of physics and chemistry. So convinced was he of this approach, that he traveled to Germany to attend lectures by Heisenberg, Sommerfeld, and Einstein—an awesome trio for the newly graduated medical student. Upon returning to the states, Hartline was offered and accepted the position of Fellow of Medical Physics by Detlev Bronk, Director of the Johnson Foundation at the University of Pennsylvania. The emerging field of biophysics had a new advocate.

In 1949 Hartline moved to Johns Hopkins University at the invitation of Bronk, who had assumed the presidency, to chair the newly created T. C. Jenkins Department of Biophysics. It was at Hopkins that Hartline first met Steve Kuffler. Steve's wife, Dr. Phyllis Kuffler, said there was an immediate oneness between them, which is not difficult to understand. Among those joining Hartline at

Hopkins, were E. F. MacNichol, Floyd Ratliff, Tsuneo Tomita, Lorrin Riggs, and Lloyd Biedler. After four years at Hopkins, again at the invitation of Bronk, the new president of The Rockefeller University, Hartline moved his laboratory to the East River campus where he remained until the end of his career. In 1967, he shared the Nobel Prize with Ragnar Granit and George Wald.

Kuffler's route to Johns Hopkins was not as direct as Hartline's. Born in Hungary in 1913 he had no formal education until after the age of ten when he was sent off to school to learn Latin and Greek. After about ten years he had narrowed his career choices to two: law or medicine. With no science background he decided for some unknown reason to study pathology in Vienna. He graduated in 1937 and left Austria the day after Hitler invaded in 1938. Landing in England he caught the next boat to Australia where he became an unpaid demonstrator in the Department of Pathology. At the prodding of Father Richard Murphy, Kuffler met John Eccles, who was looking for a tennis player. Eccles suggested he stay on in Australia, play more tennis, and work in neurophysiology, to which Kuffler responded "What's that?"

He soon found out. Within two years Kuffler developed the single fiber nerve-muscle preparation and gained immediate recognition. His wife, Phyllis later commented that Steve had beginner's luck—everything he did seemed to succeed. Steve would respond, "no luck, just work." Bernard Katz joined Eccles and Kuffler and together they became a formidable research group. In addition to research they lectured to medical students in Sydney. At one such lecture a young student noticed that Eccles and Katz were surrounded by students with questions but Kuffler was not, so she went up and asked him some questions. He must have had the right answers because they married three years later.

In 1945, after the war ended, the Kufflers left Australia for Chicago where Steve spent two years with Ralph Gerard, who at the time was developing glass microelectrodes for single-cell recording. Two years later Steve went to the Wilmer Eye Institute at Johns Hopkins and spent his first summer at the MBL. Because he was employed by an ophthalmology department, Kuffler felt obligated to do research in some aspect of vision. He picked up on Hartline's 1938 study of the receptive fields in the frog retina and extended it to the mammalian retina. His 1958 study of the cat retinal ganglion cells is a milestone in visual physiology. It set the stage for research on the mammalian vision system. Kuffler had the courage to follow his nose and a knack for identifying critical questions in neurophysiology. Each time he ventured into a different area, he seemed to strike gold and open up a new field of research for others to pursue. For example, the work he began with Katz in 1945 and later at the MBL on the crustacean neuromuscular system spawned a rich area of research on synaptic transmission which later he continued with D. D. Potter, E. A. Kravitz, J. Dudel, and others. With Potter and J. G. Nicholls, he initiated a study of the properties of glial cells, and with L. Y. and Y. N. Jan he explored the role of peptides in synaptic transmission. All were pioneering efforts that provided key insights into neural function and opened new doors for future research.

Kuffler not only had a knack for identifying critical areas of research, he also had a knack for spotting talent. While at Johns Hopkins he brought to the lab a young clinical neurologist from McGill University, David Hubel, and a recent graduate from the Karolinska, Torsten Wiesel. Hubel and Wiesel extended Kuffler's work on the cat visual system and revolutionized the field of brain research.

By 1959 Kuffler outgrew the basement laboratories at Johns Hopkins and moved with his group to the Harvard Medical School. Kuffler was convinced that the study of the nervous system required a combination of anatomy, biochemistry, and physiology and thus in 1966 created the Department of Neurobiology at Harvard. This action broke new ground including the creation of the term "Neurobiology." We all credit Steve with that word. At the MBL he created a training program in neurobiology, the precursor to the current, popular course in Neurobiology.

Both Kuffler and Hartline received many awards and honors but never felt comfortable with the formalities involved or fame that they brought. The laboratory was their first priority. Kuffler continued an active research life until the time of his death on October 23, 1980. He was at his desk in Woods Hole on that Saturday after returning from his morning swim at Stony Beach. Hartline had planned to continue his research in vision but could not, because—ironically—he was losing his sight. He occasionally traveled to New York to check on his laboratory but most of the time he lived at home with his wife Betty in Hydes, a small village north of Baltimore. He remained active, kept abreast of current research, and never missed a daily walk in the woods surrounding his home, Turtlewood. Hartline had had a series of heart problems in his later years and on March 18, 1983, died in a hospital near his home.

Hartline requested that there be no formal service or fund established in his memory. However, he said he would not object to a memorial concert as long as it included his favorite composers—the four B's. Typical of his humor, they were Bach, Beethoven, B'Mozart and SchuBert. In the summer of 1985 a public concert, organized by his family and friends, was performed at the MBL where he began his research 60 years earlier.

Colleagues of Steve established the Kuffler Fellowship Fund to support the research of young investigators here at the MBL. Thus far ten aspiring neurobiologists have been awarded Kuffler Fellowships.

Torsten Wiesel, Vincent and Brooke Astor Professor at The Rockefeller University, was a co-recipient with David H. Hubel of the Nobel Prize in Physiology or Medicine in 1981. In 1954, he received a medical degree from the Karolinska Institute in Stockholm, and in 1955 began postdoctoral work under Steve Kuffler at Johns Hopkins Medical School. In 1959 Steve's lab moved to the Harvard Medical School, where Torsten was appointed professor of neurobiology in 1968; in 1973 he took over the reins of the department as chairman. He was named Robert Winthrop Professor at Harvard in 1974. In 1983 he accepted a position at Rockefeller, where he organized the new Laboratory of Neurobiology. In 1992, Torsten became president of Rockefeller University. Torston has received many awards and honorary degrees in addition to his Nobel Prize. He is a member of the National Academy of Sciences, American Academy of Arts and Sciences, American Philosophical Society, and a Foreign Member of the Royal Society.

Literature Cited

Barlow, R. B. 1986. From string galvanometer to computer: Haldan Keffer Hartline (1903-1983). *Trends in Neurosci.* **9 (Nov/Dec):** 552-555.

McMahan, U. J. 1990. *Steve.* Sinauer Associates, Inc. Sunderland, MA.

Ratliff, F. 1990. *Haldan Keffer Hartline (1903-1983). Biographical Memoires,* ***59*** National Academy Press, Washington, DC, Pp.196-213.

Neural Mechanisms of Visual Perception: The Legacy of Hartline and Kuffler

Torsten Wiesel

The Rockefeller University

IT IS INDEED A GREAT PRIVILEGE FOR ME to give this lecture in honor of Keffer Hartline and Steve Kuffler. It is also a very moving occasion, a time for us to remember the heritage of our day-to-day scientific work and to recognize those who enlightened and helped us so much.

The work of Keffer Hartline and Steve Kuffler provided us with many of the central concepts of visual physiology. Yet what is easily lost in a summary of their achievements is a sense of their scientific styles, their quite different ways of using experiments to ask questions of fundamental importance. Hartline—like Kuffler's mentor, Bernard Katz—worked on the same system all his life. Keffer pursued the single problem of how the retina processes visual images, analyzing and attempting to model mathematically the workings of his retinal preparation in increasing quantitative detail as his career proceeded. Kuffler, on the other hand, spent only a few years—though they were immensely fertile—on retinal physiology. His life-long interest was uncovering the mechanism of synaptic transmission. He approached this problem from a variety of angles, using a variety of systems, turning his efforts to whatever preparation seemed to hold the next piece to the larger puzzle of how nerve cells talk to each other. One might compare Hartline to a novelist who wrote about the same city at progressively deeper levels, Kuffler to a novelist who pursued the same theme in many different settings. Science, like art, is a medium through which a spectrum of individual styles may find expression.

I want to begin by discussing Hartline's pioneering studies in retinal physiology. As we all know, light enters the cornea and passes through the lens to be absorbed by the phototoreceptors, the initial step in visual processing. The photoreceptors communicate with second-order retinal neurons (bipolar cells, horizontal cells, amacrine cells), which forward visual signals to retinal ganglion cells. These retinal neurons make up the machinery that processes visual information before it is sent to the brain via the optic nerve—the retinal ganglion cells' axons. Sheer numbers tell the story: there are one hundred million photoreceptors and only one million optic nerve axons, indicating that the retina must extract and compress information from the image. The advance made by Hartline in the 1930s was to record from the axons of single retinal ganglion cells.

Hartline's idea of recording from single retinal cells came from Lord Adrian, who had developed a technique for recording from the optic nerve of the eel (Adrian and Matthews, 1927). With this method Adrian found that light evoked mass discharges of nerve impulses. Knowing what Adrian and his colleagues Bronk and Zotterman had accomplished in their studies of single peripheral nerve fibers, Hartline developed his technique of dissecting single nerve fibers from the surface of the frog's retina (Hartline, 1938). He teased these nerves apart and placed a single fiber on a hook-recording electrode. Then with amplifiers and an ink writer, he could record the responses to light. [A few years later, Ragnar Granit, along with Gunnar Svaetichin, in Sweden, with whom Hartline shared the Nobel Prize in 1967, developed an electrode that could record signals from the cell body of a neuron rather than from its dissected fiber (Granit and Svaetichin, 1939). The electrode was an insulated fine wire that was inserted into retinal tissue.]

Figure 1 (from Hartline, 1938) shows a fiber that gives a strong response when the light goes on. Another fiber gives a strong response at both on and at off. Still another does not respond at on, only at off. These then are the classical "on," "on-off," and "off" types of ganglion cells. This was Hartline's first observation. The second was that the response could only be evoked by illuminating a limited area of the retina which he called the "receptive field." A cell receives information only from its nearest neighbors, not from more distant cells, allowing it to "see" only a limited segment of the visual field. The receptive field properties are based on the excitatory and inhibitory connections that give rise to "on" and "off" types of interactions.

Hartline defined the receptive field in 1938. World War II began the following year, and it was not until 1952-53 that Horace Barlow in England made the next major advance in retinal physiology (Barlow, 1953). He found that if he stimulated an on=off retinal ganglion cell of the frog in the center of its receptive field, he could record both an "on" response and an "off" response. But if he stimulated outside the area Hartline had characterized as the receptive field, the cell's response was inhibited. This is a very important observation, and it came at about the same time as Steve Kuffler's analysis of the receptive fields of retinal ganglion cells in the cat, which I'll discuss shortly.

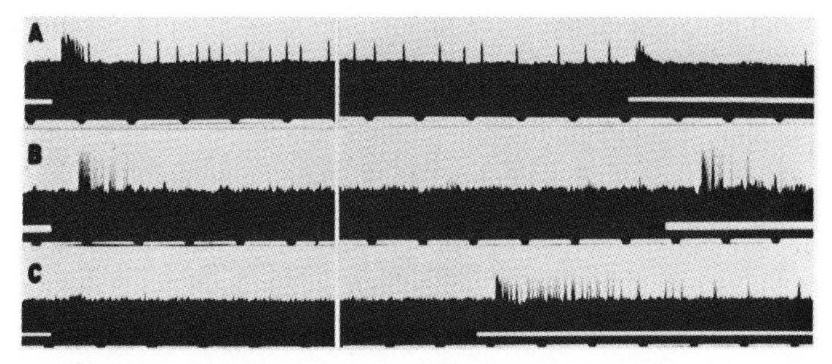

Figure 1. *Hartline recordings from three optic nerve fibers in the frog retina. (A) "On" cell responds to the onset of light. (B) "On-Off" cell responds to both the onset and cessation of light. (C) "Off" cell responds only to the cessation of light.*

I recommend that students read the 1953 paper by Horace Barlow. There he discusses how an "on-off" cell, because of its inhibitory surround, is well-suited to respond to a small dark object against a white background. Through its wiring to other retinal elements the cell has become a "bug detector." To my knowledge, Barlow was the first person to consider the functional significance of such cells in the retina. Later in the decade, Jerry Lettvin and Umberto Maturana published their famous paper titled, "What the Frogs Eye Tells the Frog's Brain" (Lettvin *et al.*, 1959). I also recommend reading this very entertaining and interesting paper. In it Lettvin and

Maturana describe a more complex set of response properties of frog retinal cells and classify the cells as interested in dimming, edges, curvature, etc. By 1960 it was clear that many of the features of the visual world that are of interest to humans are also encoded by the frog's eye.

I was particularly close to Steve Kuffler. He was my mentor scientifically as well as personally. Bob Barlow has alluded to the fact that both Hartline and Kuffler disliked pomposity and self-aggrandizement. Despite all the accolades they received, they remained bench scientists above all. Steve, and I am certain the same was true of Keffer, did not respect people who thought that there were other things more important than doing science. Steve loved to work at the Marine Biological Laboratory every summer, where he was one of the central figures sought out by students and colleagues, for whom he became a role model. Steve certainly was a role model for me, and I often feel guilty when I am away from the lab too much—but it happens, particularly lately.

Stephen Kuffler received his research training in John Eccle's laboratory in Sidney, Australia, where he became a close friend and apprentice of Bernard Katz. After World War II, Kuffler went first to Ralph Gerard's laboratory at the University of Chicago and then to the Wilmer Institute, an ophthalmology institute at Johns Hopkins University, where, among a range of different projects, he carried out his classic study of cat retinal ganglion cells. After moving to Harvard Medical School from Johns Hopkins in 1959, he assembled the group of young neuroscientists shown in Figure 2. Steve is in the middle with David Hubel, Edwin Fursphan, David Potter, and myself. Ed Kravitz was adopted a little bit later—after about ten years, our family had grown.

Before arriving at Hopkins, Steve was interested primarily in the muscle spindle and in the neuromuscular junction. Because he was in an institute of ophthalmology (his lab was located in the Wilmer Institute's basement), however, he felt he should do research in vision. He spent his first two to three years at the Wilmer trying to develop, with the help of Sam Talbot, a method for recording responses of single retinal cells in the cat to small spots of light. In his now famous 1953 paper, published in the *Journal of Neurophysiology*, Kuffler reported his findings on the receptive fields of cat retinal cells. Figure 3, which is from that paper, shows that small spots of light gave "on" responses in the center of the receptive field and "off" responses in the

Figure 2. *The Harvard group in 1959. Back row: Ed Furshpan, Steve Kuffler, and David Hubel. Front row: Dave Potter, Ed Kravitz, and Torsten Wiesel. Photograph by J. Gagliardi.*

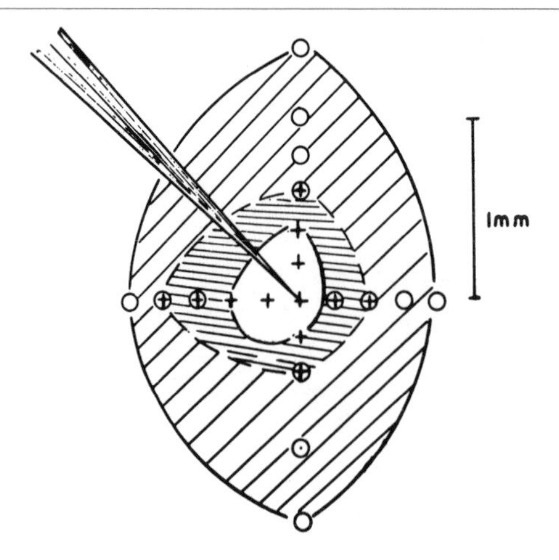

Figure 3. *Kuffler recording from cat retinal ganglion cell. The center of the receptive field (where the electrode tip is located) gives an "on" response to light; the surround (diagonal hatches) gives "off" responses. The responses are "on-off" in an intermediate zone (horizontal hatches).*

periphery. Kuffler's results correspond to some extent with what Barlow reported for the frog retina: the cells respond in an opposing manner to light in the center and surround. One difference is that in the cat, cells respond only to "on" or "off" in the center; another difference is that Kuffler could evoke responses by illuminating the periphery alone.

Kuffler found that if, by illuminating the center of the receptive field with a spot of light, he evoked an "on" response, then illumination of the surround with an annulus caused the cell's firing to stop. The cell then responded again when the annulus was turned off. Diffuse illumination stimulating both the center and the surround evoked very weak responses. Another cell gave the opposite response: "off" in the center and "on" in the surround (Fig. 3). This is the basic knowledge on which much of the field of retinal physiology rests. It is the physiological counterpart to the knowledge of the anatomy of the retina resulting from the studies of Santiago Ramon y Cajal.

Figure 4 (left, top) shows Cajal's drawing of retinal cells. It gives a wonderful sense of the retina's complexity. Direct pathways through the retina give "on" or "off" responses in the center of the receptive field. A bipolar cell, which is a second-order neuron, sends signals from a photoreceptor cell to a ganglion cell located directly below the photoreceptor. The first extensive description of the response properties of different cell types in the retina was carried out by Werblin and Dowling (1969). The study of retinal structure and function is still quite vigorous and a host of new methods and approaches has been developed in recent years. [Just as an aside, the first clear evidence that light hyperpolarizes the photoreceptor cell membrane potential was made by Tomita, who had been a postdoctoral student in Hartline's laboratory. A student of Tomita, Kaneko was one of the first to show that bipolar cells, like ganglion cells, have a center-surround receptive field organization (Kaneko, 1970).]

The major focus of my lecture is the processing of visual information at the level of the cerebral cortex, a stage at which a major transformation of information takes place. Figure 4 (left, bottom) from Polyak's book shows the central pathways of the visual system leading from the retina. Retinal nerves terminate in the first way station, the lateral geniculate nucleus (or LGN), which sends fibers to the cortex. The retinal fibers from the left half of each eye project to the left side of the cortex, the right halves to the right side. Thus anything in the right visual field is transmitted to the left part of the brain, and vice versa.

Figure 5 (top) shows the back of the monkey brain, containing the primary visual cortex that receives information from the eyes. Projections from each half of the retina are topographically organized on the cortex in a very orderly fashion. Figure 5 (bottom) shows the section taken from the right half of the cortex. Figure 6 shows cells within such a section that have been impregnated by silver to visualize them. If all cells were stained the figure would be completely black. But fortunately for neuroanatomists, the Golgi silver method stains just a few cells, thus isolating them for scrutiny. Also shown in Figure 6 is a tungsten electrode, a tool that David Hubel developed in the 1950s before we started to work together (Hubel, 1958). A wire electrode is insulated except for the very tip, and when it is close to a nerve cell you can record its action potentials.

In 1958, when David Hubel and I started our work in Kuffler's lab on the response properties of cells in the visual cortex, we of course had read Hartline and Kuffler's original papers. Hubel and I wondered how the "on" center and "off" center signals from the retinal

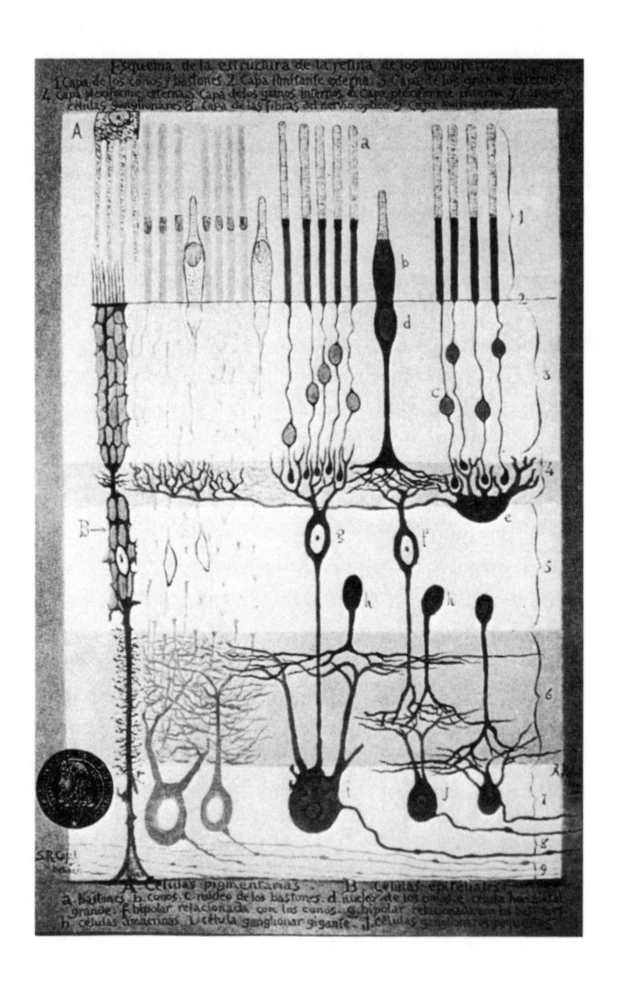

Figure 4. *(Left) Cajal picture of retinal cells.*

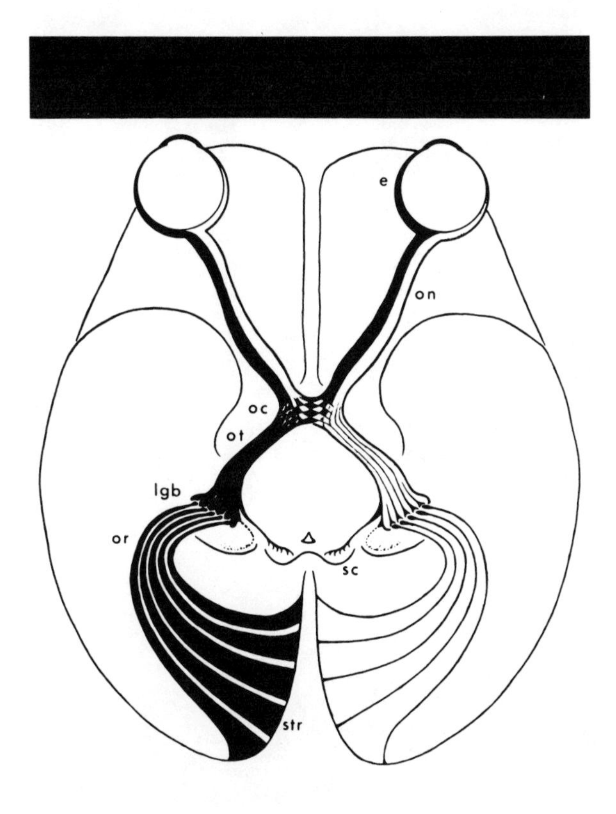

Figure 4. *(Right) Diagram of the central visual pathway from Polyak's book.*

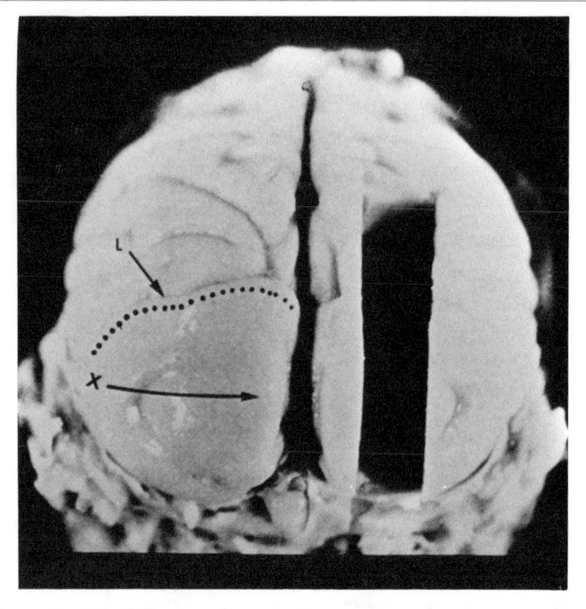

Figure 5. *(Top) Section through the macaque visual cortex. (Bottom) Macaque visual cortex seen in cross section.*

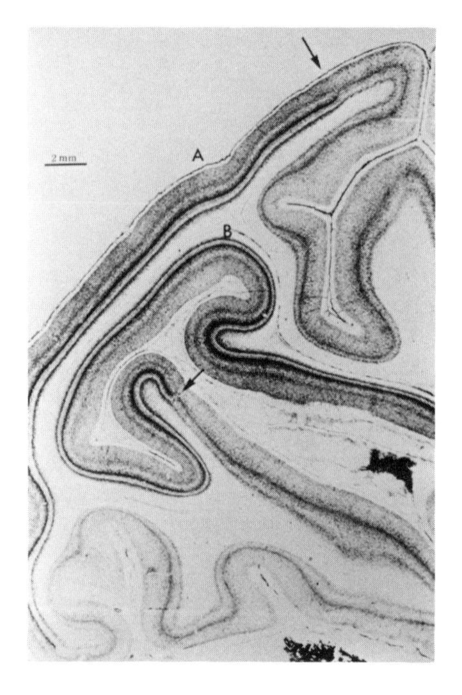

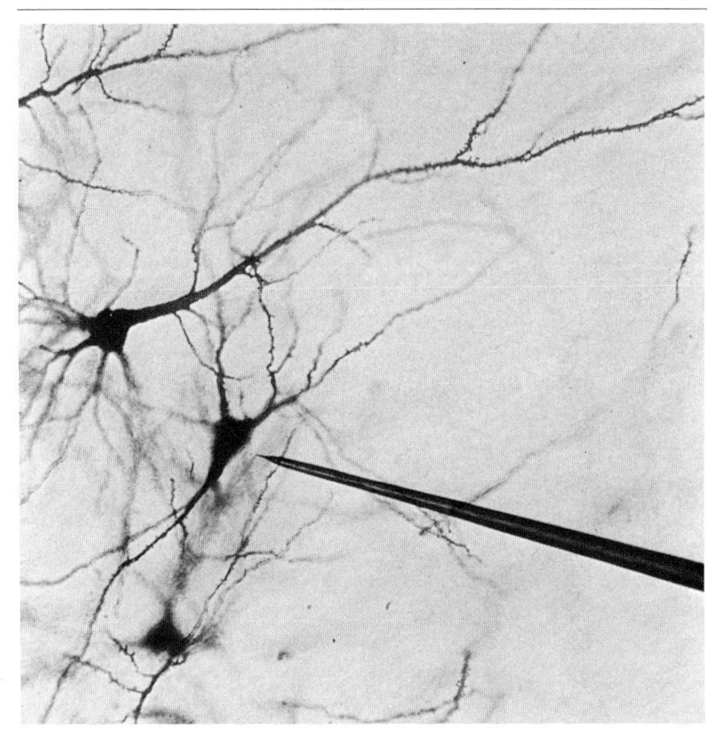

Figure 6. *The cortical cell visualized by the silver-impregnation method of Golgi. Poised above is a tungsten microelectrode used to record action potentials from the cell.*

ganglion cells were used by the central visual pathways to enable us to visually perceive objects. From their study of *Limulus*, Hartline and Ratliff pointed out that through lateral inhibition, a system of visual cells was able to exaggerate the physical contrast between areas of light and dark (Hartline and Ratliff, 1957). So it is clear that spatial antagonism is a mechanism for enhancing contrast. As you will see, the brain uses this information very cleverly.

In our initial experiments we mapped receptive fields of cortical cells using small spots of light. Small spots of light, however, gave very poor responses, and it was quite frustrating. We found it hard to make any sense of the responses. Some weeks passed before we perceived the true nature of these receptive fields. As shown in Figure 7, the region giving "on" responses is long and narrow with flanking regions of opposing polarity. This is the same type of spatial segregation of excitation and inhibition that Kuffler found in the retina. The only difference is that there were long narrow flanking regions in the cortex rather than the concentric "on-off" arrangements that Kuffler had described. The responses to spots of light were, however, quite weak and to affect the firing of the cells and to obtain an optimum response, it was necessary to use a stimulus shaped according to the cells' receptive field map—that is, a long narrow bar positioned and oriented appropriately (Hubel and Wiesel, 1962). A horizontal bar gave good responses and a verticular bar gave no response. Other orientations gave intermediate responses.

The first clue we obtained that cells responded to visual contours of specific orientations was entirely accidental, however. Sam Talbot had built a very nice ophthalmoscope for Steve Kuffler. To switch from a dark to light spot in the ophthalmoscope you had to insert an appropriate slide into a projection slot; this switching procedure caused an edge, cast by the shadow of the slide, to move across a cell's receptive field. On very rare occasions we got a very good response

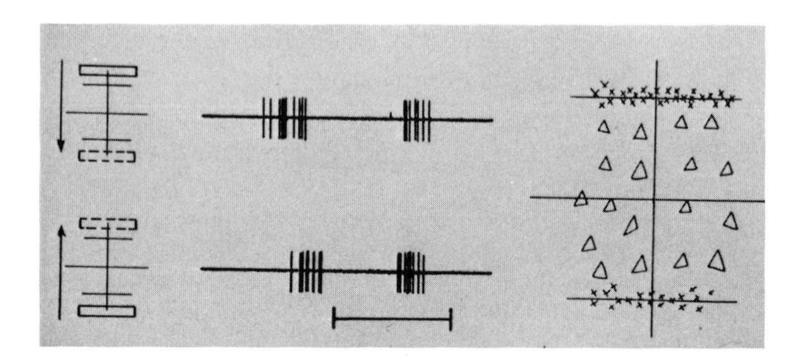

Figure 7. *(Left) Slow up-and-down movements of a horizontal slit across a simple-cell receptive field. (Middle) Burst of impulses at each crossing of the excitatory region. (Right) A receptive field map of the cell's excitatory region and flanking inhibitory zones.*

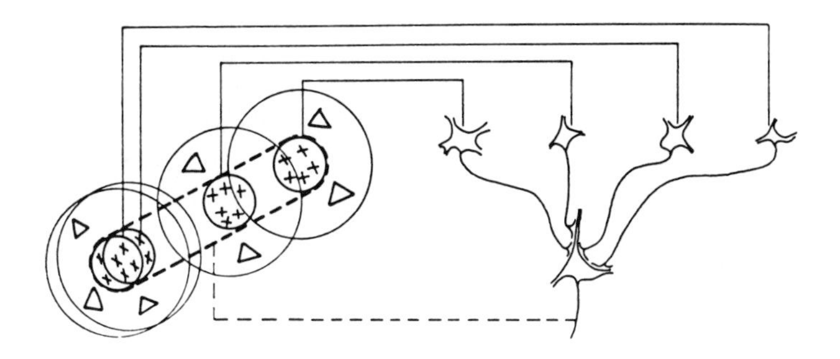

Figure 8. *The 1962 model of how the orientation specificity of a simple cell in the visual cortex may be generated by inputs from four aligned center-surround cells in the lateral geniculate.*

when we changed the slide. It took several such "accidents" before we realized the cause; it was indeed the orientation of the shadow of the slide that elicited a cortical cell's vigorous response. This insight changed our experimental approach; we threw the Talbot ophthalmoscope into a closet, resurrected an old Wilmer Institute stereotaxic apparatus from Vernon Mountcastle's laboratory next door in Physiology, and began using a projection system to stimulate the eyes of our anesthetized monkeys.

One of the first things we wanted to know was, what are the response properties of the LGN relay cells? We quickly found that they have similar response properties to those of the retina—that is, "on" centers and "off" centers. To explain the origin of orientation selectivity, we proposed that a cortical cell receives input from a row of geniculate cells that have their receptive fields along a line, as indicated in Figure 8 (Hubel and Wiesel, 1962). This model for the "simple" cell has, after years of benign neglect, received corroboration recently from the work of Michael Stryker and his colleagues at the University of California, San Francisco (Stryker *et al.*, 1990), who tested our model and found support for the construction of cortical orientation selectivity from lateral geniculate inputs.

In these experiments we also recorded from cells we named "complex," which we proposed to represent the next stage of visual processing. As is the case for simple cells, the optimal stimulus was an oriented bar of light of a certain width; but in contrast to simple cells, the precise position of the stimulus was not crucial. Thus complex cells seemed to have generalized the idea of orientation, so that the exact placement of the stimulus is no longer of prime importance. Like simple cells, complex cells have many different orientation sensitivities. We would like to think that these cells are important for analyzing form and detecting the orientation of contours. Some cells have an additional feature: they give a very good response to a short slit and no response at all to a long slit. For such cells, now called "end stopped," short bars and curved lines give good responses. Such higher-order response properties must be important if these cells are building blocks for more complex visual recognition.

Early in our work, David Hubel and I discovered interactions between the two eyes in the cortex. After we determined the receptive field properties for a neuron by stimulating one eye, we often found the same response properties to stimulation of the other eye. Such cells are receiving inputs from both retinas, and are called binocular. Their receptive fields are located at corresponding points, and when you stimulate the two eyes together you get a very strong response. Usually one eye is dominant; that is, the cell gives a stronger response to stimulation of one eye. These binocular cells, because their receptive fields lie at corresponding points, make it possible to fuse images from the two eyes together.

One important discovery, first made by Pettigrew and then by Pettigrew, Bishop, and Barlow, was that cells sometimes do not respond to stimulation of corresponding points in the retina, but instead to points of lateral disparity (Pettigrew, 1965). These cells are probably the neural basis of depth perception. The presence of such cells has been shown in the cat, and later in the monkey (Hubel and Wiesel, 1970; Poggio and Fischer, 1977). The nervous system wires these cells up in a highly precise fashion so that they have this feature of binocular depth vision with which we are all so familiar.

Such response specificities follow the legacy of Hartline and Kuffler. As one proceeds further into the central nervous system, the response properties of cells get more sophisticated, more complex, and more specific. This hierarchy of responses is clearly part of the machinery that underlies perception.

In our early studies of single neurons, Hubel and I emphasized their electrophysiological properties, but we soon realized that to understand the visual system more fully it would be necessary to use anatomical methods. In some sense, of course, our analysis of the receptive fields of single cells is an anatomical method. It is functional anatomy: understanding, by physiological methods, how cells are wired up from the photoreceptors to the cortex. In long vertical penetrations through the cortex, for instance, we found that cells lying right on top of each other had very similar response properties. If the first cell responded well to the left eye, and preferred the left eye over the right, so did all of the following cells. Similarly, if the first cell preferred a vertically oriented contour, then it was very likely that all of the cells along the path would have the same preference. (One exception was in layer-four cells of the monkey. They are very similar to geniculate cells—that is, they are not very orientation specific. Their receptive fields have a circular, symmetric organization. In the cat, on the other hand, cells in layer four in the cortex are orientation specific.)

Thus with physiological methods we found (using Mountcastle's terminology from the somatosensory cortex) a columnar cortical organization (Mountcastle, 1957). Cells along a path perpendicular to the cortical surface share common features. Most strikingly, cells with the same orientation preference form a column extending from the surface of the cortex to the white matter. These orientation columns are laid out in a highly regular way: cells located next to each other have very similar but not necessarily identical orientations; as one moves parallel to the cortex, orientation preference can shift sequentially as regularly as the hands of a clock.

This regularity of organization is also true for ocular dominance. The input to the cortex from one eye ends in a 0.5-mm-wide strip in layer four, and alternates with an equally wide strip from the other eye. So you have right eye, left eye, right eye, left eye in layer four, where cells are entirely monocular. Cells above layer four are binocular; that is, cells above a left eye strip prefer the left eye, and those above a right eye strip prefer the right eye. These are the ocular dominance columns. With anatomical methods one can show that ocular dominance is really laid out in parallel bands and not columns. It is possible to visualize these columns in the following ways: by making lesions in the LGN and looking at degenerated inputs, by using the metabolic stain 2-deoxyglucose, by transneuronal transport of radioactive amino acids from the eye through the LGN, and with an optical method worked out by Amiram Grinvald and Dan Ts'o and their colleagues

(Ts'o *et al.*,1990). This method was inspired by Larry Cohen, who developed a number of optical dyes to study neuroactivity (Cohen *et al.*, 1978). Grinvald found that you can also detect neuroactivity without using dyes. Light of a given wavelength is shone onto the cortex (with bone and dura off), and then the reflected light is analyzed. It turns out that cortical activation causes slight differences in blood flow, which can be monitored through the pattern of reflected light. So if you stimulate one eye and not the other, the light reflected by cells in one region will change relative to an adjacent region. A main source of the signal is the change from oxyhemoglobin to deoxyhemoglobin. If the light reflected from the cortex is analyzed with a CCD camera, a clear segregation of neural activity can be observed after alternate stimulation of the two eyes, enabling the ocular dominance columns to be visualized.

Grinvald and Ts'o showed that it is also possible to visualize the orientation columns in the monkey with optical imaging. When the eyes are stimulated with patterns of different orientations, the light reflected from the cortex shows a clear segregation of cells preferring different orientations as well as areas containing cells with no orientation preference. In 1972, on the basis of physiology and anatomy, Hubel and I came up with the diagram in Figure 9 showing ocular dominance columns running perpendicular to orientation columns—our so-called "ice-cube model." Organization in the visual cortex is certainly not this orderly, but the general concept nevertheless remains relevant.

In the late 1970s and early 1980s a major new development proved somewhat embarrassing to David Hubel and myself. Margaret Wong-Riley applied a cytochrome oxidase stain, which selectively stains mitochrondria-rich cells, to the primary visual cortex of the monkey. She found that the superficial layer is not uniform, as one would expect from a Nissel stain, but instead has a patchy appearance (Wong-Riley, 1979). David Hubel called these patches "blobs." They are distinct regions with higher levels of cytochrome oxidase, and thus higher levels of neural activity. Although Wong-Riley was the first to emphasize this, David Hubel and Anita Hendrickson and her colleagues extended the work to see what it meant (Horton and Hubel, 1980; Humphrey and Hendrickson, 1980). Figure 10 gives the new version of the visual cortex based on their results. The blobs are located in the superficial and deeper layers, but not in layer four. Further, they are located at the center of ocular dominance columns. Cells in the blobs tend to be monocular and show no orientation

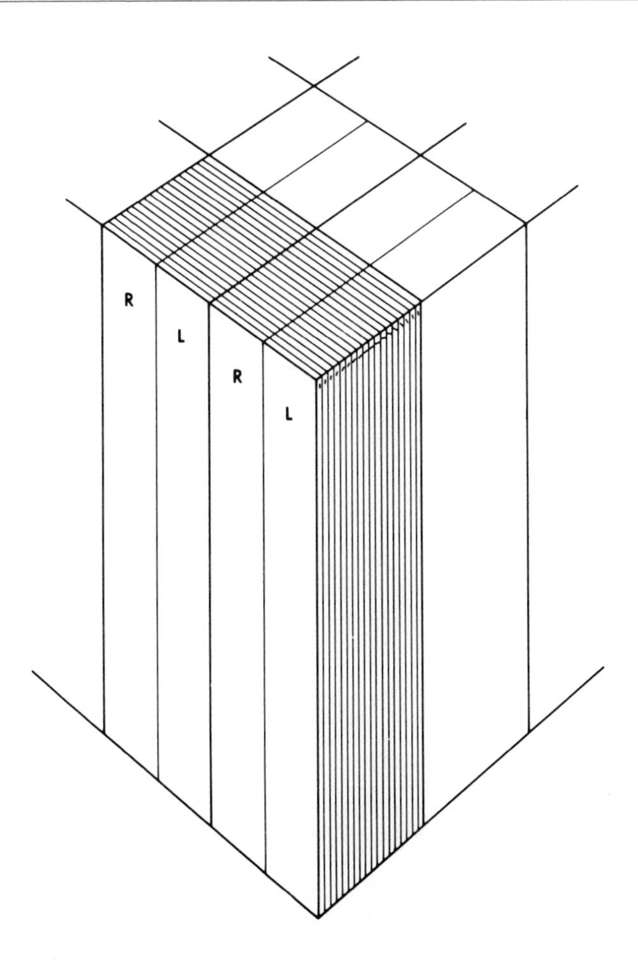

Figure 9. *The original "ice cube" model of the functional architecture of the primary visual cortex, showing the independent organization of orientation and ocular dominance columns.*

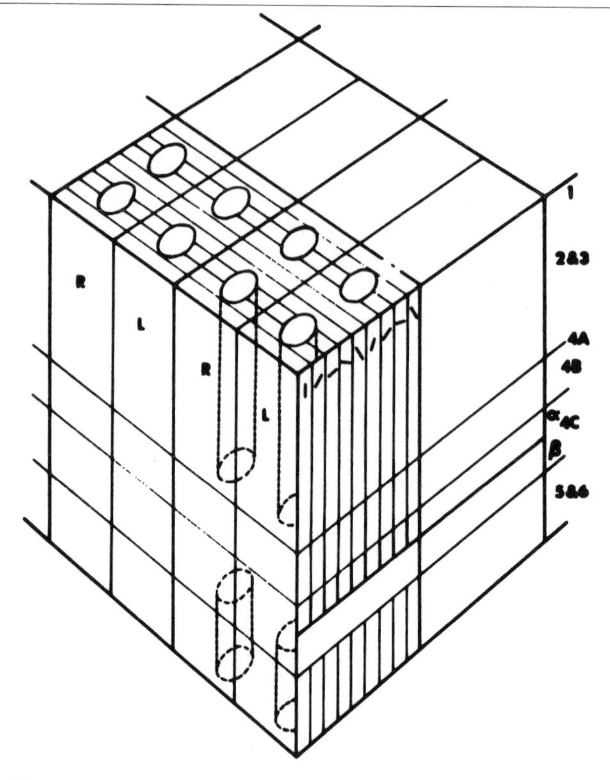

Figure 10. *An updated model of the visual cortex showing the cytochrome-oxidase "blobs."*

specificity; they are very different from the cells between the blobs. David Hubel and Margaret Livingstone showed that a large fraction of the cells, 30-50%, are color specific (Livingstone and Hubel, 1984). Dan Ts'o and Charles Gilbert recently reported that some blobs seem to have a preference for red-green coding and others for yellow-blue coding (Gilbert and Ts'o, 1988). They also showed, with me, that cells at the edge of the blobs code for both color and orientation (Ts'o, *et al.*, 1986).

Many types of visual information need to be encoded in the brain: orientation, movement, color, and binocularity. To deal with all of these different properties of the visual world, the visual cortex is segregated into a columnar-type organization for orientation and binocular interaction, and then inserts blobs at regular distances to handle color (Fig. 10).

Over the last ten years, Charles Gilbert and I have investigated the circuitry that gives cells within a column the same orientation preference (Gilbert and Wiesel, 1979; Gilbert *et al.*, 1990). The method we have used has been to record intracellularly with microelectrodes filled with horseradish peroxidase (or HRP). This method is difficult and the yield is often small, so most of our work has been done not in the monkey but in the cat. Cells in the cat visual cortex tend to be slightly larger than in the monkey cortex, but many important aspects of the organization of the visual system of the cat are similar to that of the monkey. (The main difference is that the cat has inferior color vision. Cats have orientation columns and ocular dominance columns, but, unlike the monkey, they don't have blobs; indeed, blobs are present only in primates.) We injected several hundred cells with HRP, reconstructed their processes with Camera Lucida, and made a composite with a computer graphics system. We confirmed what had been previously shown by Cajal and other anatomists of his school: that cells receiving input from the LGN project up to the superficial cells, and cells in the superficial layers in turn project to deeper layers. Cells in the middle layers are simple cells, and cells in the upper layers are complex in their features. A variety of cell types were found in a single functional column, all interconnected in a manner that allows them to respond best to the same stimulus—for instance, a vertical line.

The major advantage of combining the intracellular recording technique with HRP injections is that you get precise knowledge about both a cell's anatomy and physiology. Physiologically you can tell if cells are simple, complex, end stopped, or color coded. Anatomically, you can see the myelinated fibers extending from the cell, which is not possible with the classical Golgi method. HRP's ability to stain myelinated fibers led to a discovery by Gilbert and myself that cells not only have projections vertically up and down the laminar layers of the cortex, but over long distances tangential to the cortical surface. Figure 11 shows a computer reconstruction of the clusters of axonal terminals sent out by a layer-three cell to make contact with distant cells. The axons can run for several millimeters and the terminals always have this clustered appearance. This layer-three cell is interesting because it sends axonal projections to layer three and to layer five, and the clusters of terminals are in register both in superficial and deep layers. The periodicity of these axonal clusters, and the columnar registration of their terminals, was the first suggestion that the long-range tangential fibers connect distant cortical cells of the same orientation preference.

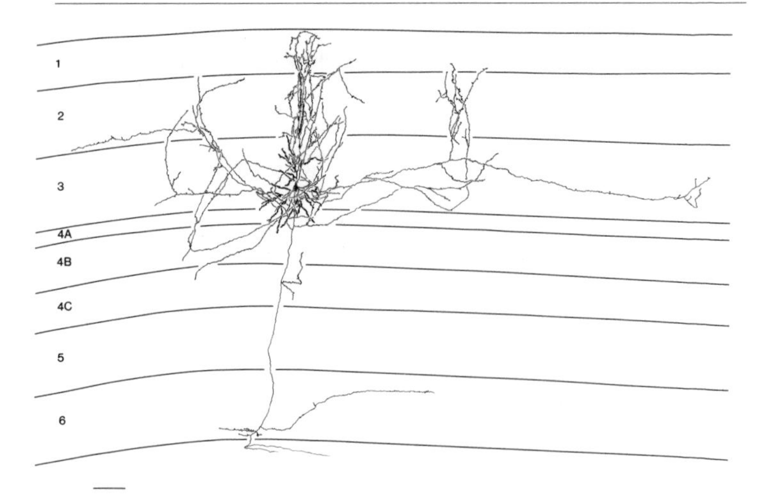

Figure 11. *Computer graphic reconstruction of a layer-three cell, showing its long-range axonal clusters.*

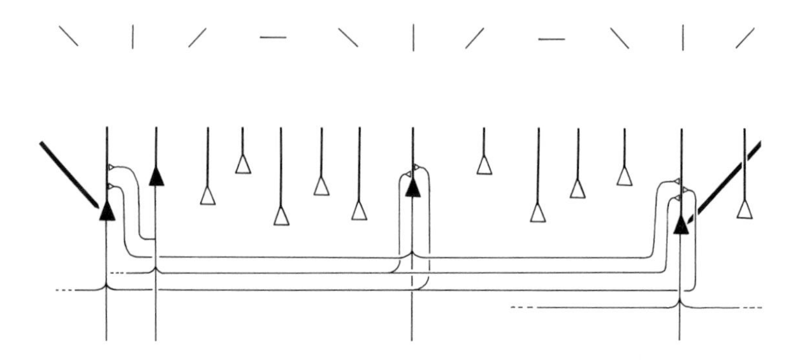

Figure 12. *A model showing columnar selectivity of horizontal connections. Axonal clusters from this cell selectively connect distant cells of the same orientation preference.*

Figure 12 gives a scheme of how cells in one vertical orientation column are connected to cells occupying other vertical orientation columns. Physiological evidence for such a scheme comes from studies in which we record from two cells simultaneously at various distances from one another and cross-correlate their responses following methods developed by Gerstein (Ts'o *et al.*, 1986). We found that there is a high probability that cells with the same orientation will have correlated responses; but cells with different orientation preferences, even if they are close together, show no correlation. Thus both anatomical and physiological results indicate that cells with similar orientation preference communicate with each other—again, an indication of the specificity of connections in the visual cortex. An important physiological point is that the classical receptive field, as defined by Hartline and Kuffler, can be expanded. Through the horizontal connections, cortical cells with receptive fields in one focal area communicate with cells at distances far outside the classical receptive field.

It is interesting to speculate about how this machinery of the brain may be related to perception. Orientation specificities in primary visual cortex (V1) seem to go to one region of V2 (secondary visual cortex), color information to another, and movement specificity to still another. The segregation of color, form, and movement information seems then to be projected to the higher areas V4 and MT (Van Essen *et al.*, 1990). In the future, we will certainly place greater emphasis on the study of the structure and function of these higher areas of the visual system. Today, following a method pioneered by David Hubel and Ed Edwards, many people are using animals with implanted electrodes with great effectiveness. Now there are a number of laboratories that study the properties of cells not only in V4 and MT, but at still higher visual stations. Results from these studies suggest that orientation selectivity is by no means the ultimate in terms of specificity: some cells in the temporal lobe of the brain seem to respond to the generalized image of a face—a stimulus consisting of two eyes, a nose, and a mouth. Other nerve cells, some researchers claim, can also tell the difference between one monkey and another. To understand how information is processed in these higher areas of the visual system, researchers are turning increasingly to behaving monkeys with implanted electrodes.

I would like to conclude with a picture of a teddy bear (Fig. 13, left). David Marr, who put this picture together, also put forward many interesting ideas about how the visual cortex may analyze

Figure 13. *(Left) A teddy bear. (Right) The image of a teddy bear analyzed according to the primary sketch model of David Marr.*

images (Marr, 1982). He suggested that the brain's first analysis is done as a kind of comic-book sketch, in which a set of cells analyzes the various contours and borders of an object. Because there is very heavy contrast around the nose and head of the bear, Marr proposed a sketch by the brain [illustrated in Fig. 13 (right)]; in these regions, contours are drawn in more thickly to indicate that the cells analyzing these contours fire more intensively. Through this sketch one immediately gets the gestalt of the object portrayed. We know from our anatomical studies that cells in the primary visual cortex are talking to each other over considerable distances. We can well imagine that even at this early level in the cortex, visual processing is more sophisticated than we might think merely by looking at a cell's classical receptive field.

Artists have often had more profound insights into the functions of the mind than scientists. When Hubel and I published our paper in 1959 showing that cortical cells are selective for line orientation, J.Z. Young sent me the cover of the *London Sunday Times* shown in Figure 14. The cover reproduces a painting by van Gogh, in which the artist constructed his self-portrait by the careful arrangement of differently oriented slashes of paint. In an accompanying note, Young said of our paper: "This is old hat."

Figure 14. *Self portrait of Vincent van Gogh anticipating modern concepts of how the primary visual cortex analyzes images.*

I hope I have given you a sense of the history of research on the visual system beginning with the pioneering work of Hartline and Kuffler. The final test of one's knowledge of the visual system is whether you can build a model that carries out all of the functions of the normal system. Then you can say, "Now I understand." We are very far from that level of understanding of the primary visual cortex, but we certainly have started to dig into it. I wish I had another lifetime.

Literature Cited

Adrian, E. D., and R. Matthews. 1927. The action of light on the eye. Part I. *J. Physiol.* **63:** 378-414.

Barlow, H. B. 1953. Summation and inhibition in the frog's retina. *J. Physiol.* **119**: 69-88.

Cohen, L. B., B. M. Salzberg, and A. Grinvald. 1978. Optical methods for monitoring neuron activity. *Annu. Rev. Neurosci.* **1**: 171-182.

Gilbert, C. D., and D. Y. Ts'o. 1988. The organization of chromatic and spatial interactions in the primate striate cortex. *J. Neurosci.* **8**: 1712-1727.

Gilbert, C. D., and T. N. Wiesel. 1979. Morphology and intracortical projections of functionally characterised neurones in the cat visual cortex. *Nature* **280**: 120-125.

Gilbert, C. D., J. A. Hirsch, and T. N. Wiesel. 1990. Lateral interactions in visual cortex. *Cold Spring Harbor Symp. Quant. Biol.* **55**: 663.

Granit, R., and G. Svaetichin. 1939. Principles and technique of the electrophysiological analysis of colour reception with the aid of microelectrodes. *Upsala Lakaref. Forhandl.* **45**: 161.

Hartline, H. K. 1938. The response of single optic nerve fibers of the vertebrate eye to illumination of the retina. *Am. J. Physiol.* **121**: 400-415.

Hartline, H. K., and F. Ratliff. 1957. Inhibitory interaction of receptor units in the eye of *Limulus*. *J. Gen. Physiol.* **40**: 357-376.

Horton, J. C., and D. H. Hubel. 1980. Cytochrome oxidase stain preferentially labels intersection of ocular dominance and vertical orientation columns in macaque striate cortex. *Soc. Neurosci. Abstr.* **6**: 315.

Hubel, D. H. 1958. Cortical unit responses to visual stimuli in nonanesthetized cats. *Am. J. Ophthal.* **46**: 110.

Hubel, D. H., and T. N. Wiesel. 1959. Receptive fields of single neurones in the cat's striate cortex. *J. Physiol.* **148**: 574-591.

Hubel, D. H., and T. N. Wiesel. 1962. Receptive fields, binocular interaction and functional architecture in the cat's visual cortex. *J. Physiol.* **60:** 106-154.

Hubel, D. H., and T. N. Wiesel. 1970. Cells sensitive to binocular depth in area 18 of macaque monkey cortex. *Nature* **225:** 41.

Humphrey, A. L., and A. E. Hendrickson. 1980. Radial zones of high metabolic activity in squirrel monkey striate cortex. *Soc. Neurosci. Abstr.* **6:** 315.

Kaneko, A. 1970 Physiological and morphological identification of horizontal, bipolar and amacrine cells in goldfish retina. *J. Physiol.* **207**: 623-633.

Kuffler, S. W. 1953. Discharge patterns and functional organization of mammalian retina. *J. Neurophysiol.* **16:** 37-68.

Lettvin, J. Y., H. R. Maturana, W. S. McCulloch, and W. H. Pitts. 1959. What the frog's eye tells the frog's brain. *Proc. IRE.* **47**: 1940-1951.

Livingstone, M. S., and D. H. Hubel. 1984. Anatomy and physiology of a color system in the primate visual cortex. *J. Neurosci.* **4**: 309-356.

Marr, D. 1982. *Vision: A Computational Investigation into the Human Representation and Processing of Visual Information.* W.H. Freeman, San Francisco, CA.

Mountcastle, V. B. 1957. Modality and topographic properties of single neurons of cat's somatic sensory cortex. *J. Neurophysiol.* **20:** 408-434.

Pettigrew, J. D. 1965. Binocular interaction on single units of the striate cortex of the cat. Sc. B. (Med.) Thesis, University of Sydney, Australia.

Poggio, G.F., and B. Fischer. 1977. Binocular interaction and depth sensitivity in the striate and prestriate cortex of the behaving monkey. *J. Neurophysiol.* **40**: 1392-1405.

Stryker, M. P., B. Chapman, K. D. Miller, and K. R. Zahs. 1990. Experimental and theoretical studies of the organization of afferents to single orientation columns in visual cortex. *Cold Spring Harbor Symp. Quant. Biol.* **55**: 515.

Ts'o, D. Y., C. D. Gilbert, and T. N. Wiesel. 1986. Relationships between horizontal interactions and functional architecture in cat striate cortex as revealed by cross-correlation analysis. *J. Neurosci.* **6:** 1160-1170.

Ts'o, D. Y., R. D. Frostig, E. E. Lieke, and A. Grinvald. 1990. Functional organization of primate visual cortex revealed by high resolution optical imaging. *Science.* **249:** 417.

Van Essen, D. C., D. J. Felleman, E. A. DeYoe, J. Olavarria, and J. Knierim. 1990. Modular and hierarchical organization of extrastriate visual cortex in the macaque monkey. *Cold Spring Harbor Symp. Quant. Biol.* **55**: 679.

Werblin, F. S., and J. E. Dowling. 1969. Organization in the retina of the mudpuppy, *Necturus maçulosus. J. Neurophysiol.* **32:** 339-355.

Wong-Riley, M. T. T. 1979. Changes in the visual system of monocularly sutured or enucleated cats demonstrable with cytochrome oxidase histochemistry. *Brain Res.* **171**: 11-28.

Charles Otis Whitman
(1842 - 1910)

William Morton Wheeler
(1865 - 1937)

Analyzing the Superorganism: The Legacy of Whitman and Wheeler

Introduction by Jelle Atema

In classical music, when we refer to the three Bs everyone knows we mean Bach, Beethoven, and Brahms. In behavioral biology, we refer to the three Ws: Whitman, Wheeler, and Wilson. Edward Osborne Wilson is one of the foremost biologists of our time and for many years a friend of the MBL. Similarly, his two predecessors, **Charles Otis Whitman** and Whitman's student **William Morton Wheeler**, were MBL faculty with profound influence on behavioral biology. In addition, Whitman was critical in the shaping of the MBL in its formative years.

Whitman was the first director of the MBL. It was generally recognized by his contemporaries and by those who followed that Whitman's persuasive idealism during the first 15 years of shaky and uncertain MBL existence steered the laboratory in the direction we now recognize—over 100 years later—as eminently successful: institutional independence and freedom of inquiry. Whitman recognized the need for specialization in scientific research when many of his peers still adhered to the old ideal of knowing and teaching all of biology. But he also recognized the dangers of fragmentation. With specific reference to the cooperation that exists between specialized parts of an organism, he was a staunch advocate of specialization AND cooperation. E. O. Wilson has used a similar metaphor when discussing the castes of insect societies. Add to this Whitman's insistence on the mixing of teaching and research and you see the MBL we all know and love.

Naturally, Whitman did not come to his philosophy in a vacuum. Perhaps it all started in 1873 when the 30-year-old teacher in Boston's "English Highschool" participated in Louis Agassiz's Summer School at Penikese Island. After attending Penikese Island School in 1873 and 1874, Whitman left high school teaching and went to Leipzig where he received his Ph.D. under Rudolph Leukart for his work on the leech *Clepsine*. He then taught at the Imperial University of Tokyo for two years, where he became known as the "Father of Japanese zoology." Subsequently, he worked at Harvard's Museum of Comparative Zoology under Alexander Aggasiz, and directed the Allis Lake Laboratory in Milwaukee. His employer there, Chicago businessman Edward Allis, was a devoted amateur scientist, who organized this lake shore laboratory for his personal education and inquiry. He wanted to run the lab as a business, with scientific discovery as its product. Whitman opposed this idea, insisting that each worker should have absolute freedom to pursue the search for truth wherever it should lead. Not only did Whitman's ideas prevail, but he also convinced Allis to sponsor publication of the *Journal of Morphology* to be a model of scientific excellence and intellectual independence.

Early in 1888 Whitman accepted the position of MBL Director, a summer job he continued for the next 20 years. Meanwhile, he became chairman of Zoology at Clark University in 1889, and in 1892 he organized and headed the biology program in Chicago, where he remained until his sudden death in 1910.

Soon after the founding of the MBL, Whitman organized the Biological Lecture series, which were published from 1890 to 1900 and were based on "Friday Evening Lectures." His inaugural lecture (1890) was titled "Specialization and Organization: Composition Principles of All Progress—The Most Important Need of American Biology." This lecture was something of a manifesto for the MBL. Its principles were to be defended several times over the subsequent decades, which established the MBL as one of the foremost nurseries for American Biology.

While the University of Chicago repeatedly failed to provide a marine laboratory or an island experimental farm as Whitman urged, he established a research facility in his own backyard. There, Whitman set up a pigeon colony and devoted himself to the study of their life history, evolution, and behavior. Traveling—with pigeons—back and forth to Woods Hole for several summers proved expensive and tiring, but Whitman had committed himself to studying pigeons to answer questions on the evolution of behavior patterns. This commitment brought the study of instinct and behavior to the MBL in the late 1890s as it had not been before and has not been since.

In August 1898, Whitman delivered what is now considered his most influential paper at one of the Friday Evening Lectures. It was titled "Animal Behavior," and began with a quote from Darwin, who said in *The Origin of Species*: "'Natura non facit saltum' is applicable to instincts as well as to corporeal structure." Under this motto, Whitman discussed eloquently the principles of modern ethology, which were hardly heard of again until, in the 1950s, the European ethologists and subsequent Nobel laureates Konrad Lorenz and Niko Tinbergen redeveloped and elaborated these themes. Tinbergen's famous "Four Questions" state that we can only begin to understand the performance of a behavior pattern if we know its phylogeny, its ontogeny, its plasticity, and its underlying physiological mechanisms. Whereas Jacques Loeb focused increasingly on physico-chemical mechanisms, Whitman and others maintained a multidisciplinary perspective for biology. The following quote from Whitman's lecture on behavior may provide a glimpse of the discussions that captivated biologists and lay people in Whitman's day. After a description of his careful observations of feeding behavior in the salamander, *Necturus*, Whitman wrote on the "Origins and Meaning of Behavior":

"We have taken a very important step in our study when we have ascertained that behavior, which at first sight appeared to owe its purposive character to intelligence, cannot possibly be so explained, but must depend largely, at least, upon the mechanism of organization. The *origin* and *meaning* of the behavior antedate all individual acquisitions and form part of the problem of the origin and history of the organization itself. It is the first and indispensable step, without which it would be impossible to reach sound views, either as regards the particular behavior or the difficult question of the relation of instincts to habit and intelligence. If the problem is *not simplified*, its nature is better defined and its perspective is relieved of many a myth that might otherwise obscure our vision. We see at once that the behavior does not stand for a simple and primary adaptation of a pre-existing mechanism to a special need. As the necessity for food did not arise for the first time in *Necturus,* the organization adapted to securing it must be traced back to foundations evolved long in advance of the species. The retrospect stretches back to the origin of the vertebrate phylum, and, indeed, to the very beginning of genealogical lines in protozoan forms. The point of special emphasis here is that instincts are *evolved*, not *improvised*, and that their genealogy may be as complex and far-reaching as the history of their organic bases." (my italics, J.A.)

In my readings, Professor Whitman has become a friend, a much respected and trusted colleague, whose views of biology ring true and clear today after nearly a century of discovery of facts. Whitman's influence reaches us not only through his own writing but also through the legacy of his students, among them William Morton Wheeler.

Wheeler was an enthusiastic Whitman student, first at the Allis Lake Laboratory, next at Clark University where he received his Ph.D. in 1892 (under Whitman), and then at Chicago. In 1898 Wheeler helped Whitman organize the *Zoological Bulletin*, which two years later changed its name to *The Biological Bulletin* with editorial offices in Woods Hole, where it still flourishes. The next year he accepted an appointment to direct the Zoology Department at the University of Texas in Austin, then a frontier country and a cultural desert. But animal life was abundant and one day in a dry stream bed, lamenting the lack of tools and facilities he had been used to, Wheeler saw a file of ants march through a sand trail. It appears that at that moment he "thought what nobody had thought, while seeing what everyone had seen" (to quote Albert Szent-Györgyi), and he proceeded to become a world authority on ants.

Wheeler left Texas in 1903 to become curator at the American Museum of Natural History, then under the direction of his old MBL colleague Herman Bumpus. Four years later he joined the Harvard faculty at the Bussey Institution, which in 1915 became the Graduate School of Applied Biology with Wheeler as Dean. Wheeler was a prolific writer, publishing several books and often over 10 papers per year. Many dealt with ant biology, taxonomy, and geography. In 1928 he wrote *The Social Insects: Their Origin and Evolution*; the next time a book on this subject appeared in the English language was 1971, half a century later, when E. O. Wilson published *Insect Societies*.

Although Wheeler's association with the MBL was brief, he is an important link between C. O. Whitman and E. O. Wilson for more than the simple reason that Wheeler and Wilson are both world authorities on ants, or the fact that Wilson sits at Wheeler's old desk at Harvard. In 1911, Wheeler gave a lecture at the MBL in which he apologized for being an old-fashioned student of insect life rather than an experimentalist. This "apology" was probably directed at Jacques Loeb and his enormous influence on biological thought and method. Although Loeb and Wheeler remained friendly, their views of biology became diametrically opposed. Wheeler, as did Whitman, Jennings,

Sherrington, and Von Uexkull, considered Loeb's tropisms only a small part of an organism's behavior. Loeb's reductionism proved extremely persuasive as is all too apparent today. Modern biology has sliced the processes of life down to their component macromolecules. This would not have surprised Loeb. The great question for today's biology is to learn how life *emerges* from its component molecules. And here is where natural history in its broadest sense remains the touchstone of success. The natural history of an individual organism, of a single cell, of an assemblage of cells, and of assemblages of organisms will always remain the ultimate standard against which we must measure the success of attempts at synthesizing life—attempts that will take place until well into the twenty-first century.

E. O. Wilson is such a synthesizer. His research, as was Whitman's and Wheeler's, is founded on a solid first-hand knowledge of the natural history of his primary study subjects: ants. He has sliced their behavior and social organization into their component parts. And from there he has re-synthesized their natural biology. In 1975 he organized our current knowledge of animal social behavior in perhaps his most famous book *Sociobiology*, in which he builds an impressive data base for Darwin's insights and observations on the phylogeny and evolution of behavior, as championed by Whitman and Wheeler.

Wilson is the Baird Professor of Science and Curator in Entomology at Harvard University. Born in Birmingham, Alabama in 1929, he earned a B.S. and an M.S. at the University of Alabama before coming to Harvard to commence studies toward his Ph.D. degree in 1951. At that time in his career—well ahead of Wheeler—he was already a recognized authority on ants. He joined the Harvard faculty in 1956 and he has served there ever since. His best-known books, which reflect the different stages in his research interests during the past 30 years, include *The Theory of Island Biogeography* (with Robert MacArthur, 1967), *The Insect Societies* (1971), *Sociobiology* (1975), *On Human Nature* (1978), *Biophilia* (1984), *The Ants* (with Bert Hölldobler, 1990), and *The Diversity of Life* (1992). Wilson's research, now devoted primarily to ants, is conducted in the laboratory and library, as well as in the field in Brazil and Costa Rica.

In his centenary essay on behavior, Wilson presents the modern version of the themes of ethology and sociobiology, with special reference to the social insects. Using examples from his own research conducted partly at the MBL, he shows how the most complex ant colonies are knit together through the interaction of ten or fewer castes communicating with one another using primarily ten to twenty classes of pheromones (chemical signals produced by exocrine glands located throughout the ant's body). As did Whitman when he compared scientific specialization and cooperation with the coordination between body parts, Wilson examines parallels between sociogenesis (the assembly of insect society during the growth of the colony), and morphogenesis (the assembly of tissues and organs during the growth of the organism).

E. O. Wilson has been the recipient not only of many honors but also of much abuse as a result of the political climate of the sixties and seventies, when free will had to be victorious over genetic determination. Wilson weathered this storm and, like Beethoven, nourished his soul at Nature's breast. He has said [1] "My ultimate retreat is the natural world through which we are privileged to travel an endless Magellanic voyage." His essay allows us the privilege to travel this world in our minds with Wilson as our guide.

[1] **Dewsbury, D. A. 1985.** Pp. 464-484 in *Leaders in the Study of Animal Behavior: Autobiographical Perspectives*, D. A. Dewsbury, ed. Bucknell University Press, Lewisburg, PA.

ANALYZING THE SUPERORGANISM: THE LEGACY OF WHITMAN AND WHEELER

EDWARD O. WILSON

Harvard University

MUCH OF THE HISTORY OF BIOLOGY can be expressed metaphorically as a dynamic tension between unit and aggregate, between reduction and holism. An equilibrium in this tension is neither possible nor desirable. As large patterns emerge, ambitious hard-science reductionists set out to dissolve them with nonconforming new data. Conversely, whenever empirical researchers discover enough new nonconforming phenomena to create chaos, synthesizers move in to restore order. In tandem the two kinds of endeavors nudge the discipline forward.

So it has been with the superorganism concept. In the time of Whitman and Wheeler it was popular to compare colonies of animals with organisms, the analogs of the body parts being the specialized zooids or workers and the analogs of body processes being the communication and exchange of nutrients among the colony members. In his classic article "The ant-colony as an organism" (1911) and later in *The Social Insects: Their Origin and Evolution* (1928), Wheeler stressed the analogies between insect colonies and conventional organisms. Both behave as a unit, both undergo species-specific cycles of growth and reproduction that are clearly adaptive, and both are differentiated into germ plasm (the queens and males in the case of the colonies) and somatic units (workers in the colonies). The exchange of liquid food among colony members is the equivalent of circulation, the soldiers of ants and termites are like lymphocytes, and so forth.

The era of analogy and holism was succeeded during the 1950s through the 1970s by a period of intense analysis, during which details of chemical communication and division of labor were worked out in many colonial species, especially the social insects. Emphasis was then placed on the chemistry of the pheromones, the exocrine sites of manufacture of pheromones and nutrient materials used in exchange, the determination of castes, kin selection and kin recognition, and similar more purely empirical aspects of sociobiological research.

This phase of research continues today with increasing momentum, but it is accompanied by a more sophisticated interpretive program that collects the data into evolutionary scenarios and helps to guide research in new directions.

The Two Evolutionary Routes to the Organism

An example of the product of the more holistic approach is the recognition of the two routes to the creation of complex organisms. The pathway that is more dominant, in the sense of generating the greater variety and biomass, is the elaboration of new tissues and organs within the bodies of preexisting metazoan organisms. An eye is added, or a poison gland, or some other novel structure, by either the diversification of a previously existing structure or the *de novo* assembly of the organ from a relatively undifferentiated structure such as epithelium. Some of the colonial coelenterates have followed the less frequent but more spectacular route. In essence, the specialization and cohesion of individual colony members, *i.e.*, the zooids, have advanced so much as to render the colony indistinguishable from a complex metazoan organism. The process can be inferred from the existence of species occupying a sequence of evolutionary grades that appear logically to represent steps to the advanced superorganism.

This full sequence is represented by the living species of corals alone (Bayer, 1973). A few members of the order Stolonifera have colonies consisting of nothing more than independent zooids connected plant-like at their base by a stolon. New zooids sprout seriatim from the growing end of the stolon. Other stoloniferans form colonies when zooids sprout from the body walls of preexisting zooids, and then give rise to granddaughter zooids, creating a tree-like structure that increases in density from bottom to top. Corals of the order Alcyonacea display more advanced stages in integration. The zooids of some become differentiated into two classes of specialists: elementary autozooids that eat, digest, and distribute food to the remainder of the colony, and more derived siphonozooids that contain the reproductive organs and circulate water by means of large ciliated grooves running the length of the pharynx. The colony of one alcyonacean, *Bathyalcyon robustum*, is the ultimate form, comprising a single giant autozooid in whose body wall reside large numbers of daughter siphonozooids.

In effect the *Bathyalcyon* colony is an organism, and the organism is a colony. The paradox is worth citing to illustrate the double route that has been followed in evolution to the organism, and the basic soundness of the superorganism concept. The usefulness of the concept has been enhanced by the development of the endosymbiotic theory of the origin of the eukaryotes (Margulis, 1981). This view, which is supported by substantial but not yet conclusive evidence, holds that eukaryotic cells originated well over a billion years ago, in late Proterozoic times, by the incorporation of prokaryotic cells one into the other and a subsequent specialization of host and invader. Thus chloroplasts and mitochondria might have begun as independent prokaryotic species that were ingested by other prokaryotes, survived, and evolved into a mutually beneficial symbiotic relationship with the hosts. The endosymbiotic theory has gained greatly in plausibility from the existence of *Myxotricha paradoxa*, a protist living symbiotically in the guts of certain Australian termites. The "flagella" of the protist are actually spirochete bacteria, which are attached at regular intervals on the surface of the *Myxotricha* and beat rhythmically to carry it along. Inside the *Myxotricha* body are bacteria of a third kind believed to assist in the digestion of wood. In terms of levels of the organism-superorganism conception of life, *Myxotricha* offers a complete tableau: the termite colony comprises tightly organized termites containing symbiotic supercells comprising *Myxotricha* and its endosymbionts.

The Ant Colony as a Superorganism

In the case of social insects the accumulation of new empirical knowledge has rendered the superorganism concept more robust; the concept in turn has become a more effective stimulus to experimental research. In 1927 Wheeler first pointed out certain advantages of the insect colony over ordinary organisms in the study of biological organization. The colony can be torn apart into age or size groups, studied in fragments, then reassembled into the original whole with no harm done. The next day the same colony can be vivisected in still another way, then restored to the original state—and so on. The procedure has two enormous advantages. It is of course quick and technically easy, compared to analogous experiments on organisms.

But it also provides a built-in control of some elegance: by using the same colony repeatedly, we eliminate variance due to genetic differences and idiosyncrasies of development and experience prior to the experiment.

I used these advantages when investigating the efficiency of the foraging caste of leafcutter ants in the genus *Atta*, which inhabit the New World tropics (Wilson, 1980, 1983a). Workers of the species on which I have worked, *A. cephalotes* and *A. sexdens*, vary in size from tiny individuals with head widths of 0.8 mm to relative giants with head widths of 5.0 mm or more. Workers in the size range 1.8-2.4 mm are strongly specialized for foraging and harvesting fresh leaves and other vegetation. Although they make up less than 10% of the adult nest population, they contribute more than 90% of the cutting and harvesting. The modal group, displaying peak activity, have head widths of 2.0-2.4 mm. Is this the most advantageous deployment of labor resources with respect to the growth and reproduction of the colony as a whole? In other words, is it adaptive in the Darwinian sense, where the colony, or more precisely the progenitrix queen, is the unit of selection? I approached this question by first drawing up a list of all of the selection forces that could be inferred from existing knowledge of the natural history of *Atta*. They included various aspects of predation and competition, as well as the advantage accruing from maximum net energetic yield. With reasonable assurance I was able to eliminate predation and competition, by noting that these pressures are more readily met by other castes. For example, the large soldiers are clearly specialized for combat. This left energetic yield. My reasoning was as follows: if I were to test for net energetic yield as a function of worker size and found that cutting and harvesting could be more efficiently accomplished by workers smaller or larger than 1.8-2.4 mm, the adaptive hypothesis of division of labor in the species would be imperiled.

This last point concerning the vulnerability of the adaptation hypotheses is worth stressing, because one occasionally encounters the claim that adaptation is a tautological proposition that cannot be tested decisively. Optimization, it is further averred, is a meaningless idea because the criterion of the optimum is dependent on the observer's concept of a goal. But if optimization criteria are based on known or inferred selection pressures, adaptation hypotheses can be put to the test.

I tested the adaptation hypothesis in leaf harvesting by the "pseudomutant" method. Each day the foraging *Atta* and their attendants, whose head width ranged from 1.2 mm to over 4 mm, were allowed into an arena containing fresh leaves. As the column pressed through the exit, I removed all but a particular size class, such as 1.2 mm, or 1.4 mm, or 2.8 mm, and any other size randomly chosen for that day. The colony was thus transformed into a pseudomutant, identical in all respects to the "normal" colony (itself on other days). The experiment was rather like testing the manual efficiency of the human hand by painlessly removing or adding a finger at the start of the day, measuring its performance, and then painlessly restoring the five-fingered condition at the end of the day. The net energetic yield of each pseudomutant, that is, each size class working alone, was measured in amount of vegetation harvested both (1) per gram weight of worker weight, an index of the construction cost, and (2) per unit volume of oxygen consumed, an index of maintenance cost. The highest net yield proved to be that obtained by workers with a head width 2.0-2.2 mm, the size group in fact committed by the *Atta* colony to foraging.

An even more persuasive test of efficiency in colonial organization is provided by the pattern of colony ontogeny in *Atta cephalotes*, that is, the different frequencies of various size and age groups during colony growth from the first brood raised by the young queen to the fully mature colony producing new virgin queens. I have conjectured that the course of this ontogeny is adaptive and is in fact regulated by birth and survivorship schedules and feedback controls that constitute an adaptive demography (Wilson, 1968, 1971, 1983b; Oster and Wilson, 1978).

Consider in this light the life-and-death problem confronted by the founding *Atta* queen. She leaves the home colony on a nuptial flight, during which she is met and inseminated by five or more males. She drops to the ground, sheds her wings, and excavates a vertical gallery in the soil. She widens the bottom of the gallery into a single room, within which she will attempt to rear her first brood of workers while simultaneously growing a small garden of the symbiotic fungus on which the colony is later to depend. To accomplish these tasks, the queen must convert most of her energy reserves by metabolizing her abdominal fat body and the huge but now-useless wing muscles. She is in a race against time, with virtually no room for error. She is required to produce some workers with head widths of 1.6 mm or greater, because 1.6 mm is the minimum size needed to cut vegetation for the fungus garden substrate. The queen needs to keep the size as close to 1.6 mm as possible, because the dry weight of workers—hence the manufacturing cost—of the workers increases as the 2.5 power of the head width. If the queen makes the mistake of producing a single large media worker (3.0-4.6 mm) or major worker (over 4.6 mm), so much of the energy reserve will be used that few other individuals of any size can be generated in the first brood. The little colony will then starve. Similarly, the queen must produce some tiny workers with head widths of 0.8 mm or close to it, the size cohort specialized for gardening the delicate fungal strands. Finally, the queen needs to produce workers of intermediate sizes, 1.0-1.4 mm. These cohorts are the assembly line that converts fresh vegetation into the paste-like material on which the symbiotic fungus grows (see Figs. 1 and 2). Starting with the 1.4-mm cohort and progressing downward in size, the specialists sequentially chop the leaves into 1-mm-wide fragments, then pass the material on to smaller workers, who chew the fragments into balls of pulp, then pass them on to still smaller workers, who place the boluses onto the substrate, whereupon smaller workers place fungal hyphae on the newly implanted substrate. Finally, the 0.8-mm modal group takes over the care of the fungus.

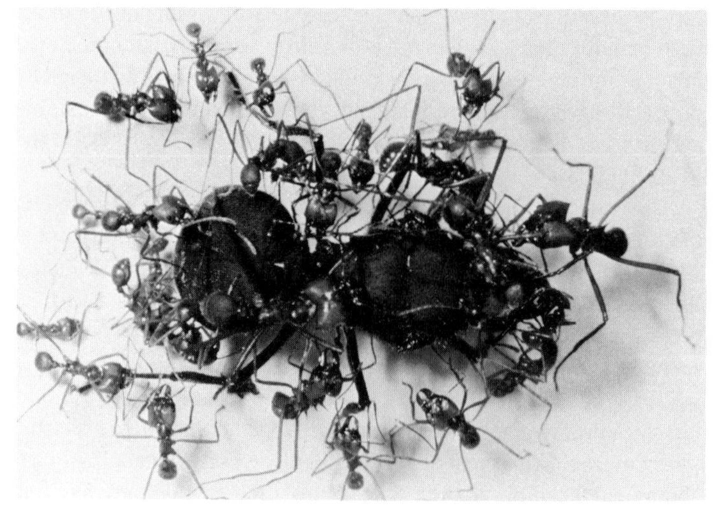

Figure 1. *A young queen of the leafcutter ant* Atta cephalotes *surrounded by daughter workers. (Photograph by Bert Hölldobler.)*

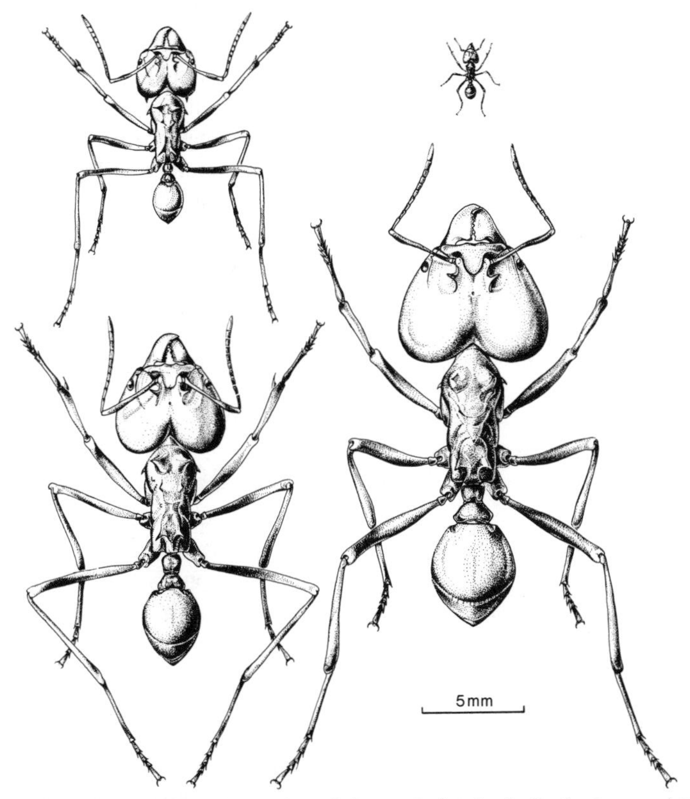

Figure 2. *The entire range of size variation in the leafcutter ant* Atta laevigatus, *comprising with the queen the full caste system; the systems of the experimental species* A. cephalotes *and* A. sexdens *are closely similar. (Reproduced with permission from* Caste and Ecology in the Social Insects. *G. F. Oster and E. O. Wilson. Princeton University Press. 1978.)*

Elaborate calculations are not needed to predict that a founding *Atta* queen must somehow rear a first crop of adult workers whose head widths range more or less continuously from 0.8 mm through 1.6 mm. She must somehow make a physiological choice, because in later stages of colony growth she has the capacity to produce workers who vary continuously from 0.8 mm through 5.4 mm or greater. And that is exactly what founding queens do. At least that is what they do in the case of *Atta cephalotes*, which I have studied most closely in this respect. They create a nearly uniform size-frequency distribution with a range of 0.8-1.6 mm, with only an occasional queen creating a 1.8-mm worker.

What is the nature of this superorganismic control? Does it come from the age of the queen and colony or from the population size of the colony? To find out, I let four colonies of *Atta cephalotes* grow for a period of 3 to 4 years, so that the populations were approximately 10,000, with major workers present (Fig. 3). I then cut them back to 236 workers each, trimming size cohorts so that the size-frequency distribution resembled that of a young colony with 236 workers. The size-frequency distribution of these "reborn" colonies resembled that of naturally young colonies. They did not approximate the distributions of colonies 3 to 4 years old. In other words, the size of the colony, not the age, determines the caste distribution. The feedback mechanism behind this remarkable control remains to be elucidated.

Feedback Loops at the Organismal Level

Although the determinants of the *Atta* caste ratios as a function of colony size remain unknown, the mechanisms of an increasing number of other colony-level controls in caste and division of labor have been discovered. In some cases, the feedback loops turn out to be slow, mediated by pheromones that act on the physiology, growth, and development of the colony members. In other cases they are relatively quick, comprising behavioral responses to special signals provided by nestmates in particular circumstances (Wilson and Hölldobler, 1988).

The ant genus *Pheidole* provides an example of slow feedback. Each species of this large, primarily tropical genus has a characteristic ratio of small-headed minor workers, which constitute the ordinary "workaday" caste, and major workers, which constitute the "soldiers." When the ratio is altered from the norm experimentally or by a natural excess of birth or mortality, the colony converges back to the original ratio within one or two worker generations, a period requiring one to three months according to species. In at least the case of *Pheidole dentata*, a species inhabiting the southern United States, even colonies differ in their equilibrial ratios, and this variation appears to have a genetic basis (Johnston and Wilson, 1984). The feedback loop establishing an equilibrium in *P. bicarinata* has been elucidated by Wheeler and Nijhout (1983, 1984). It contains an inhibiting pheromone produced by major workers that lowers the sensitivity of larvae to juvenile hormone, so that larvae surrounded by an excess of majors curtail growth and tend to mature as minors. Those present during a shortage of majors therefore receive less of the pheromone, become more sensitive to juvenile hormone, extend their period of growth, and thence metamorphose into majors.

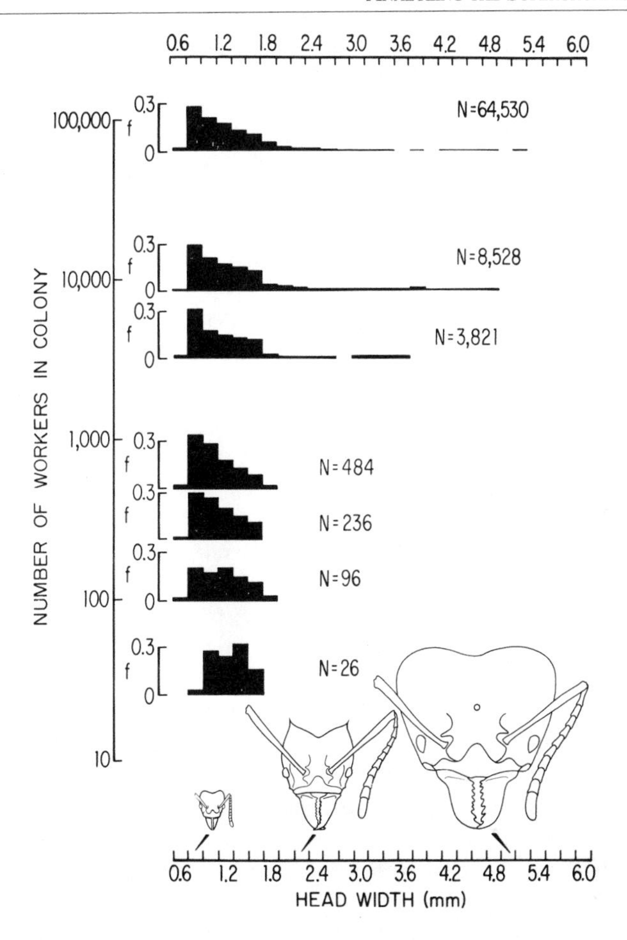

Figure 3. *Sociogenesis in colonies of the leafcutter ant* Atta cephalotes *consists partly of changes in the size-frequency distribution during colony growth. Size variation among the adult workers, also illustrated in this figure, is accompanied by large differences in head proportions caused by allometric growth of the imaginal disks during the pupal stage. Workers of different sizes are strongly specialized for labor, including the several steps required to incorporate fresh leaves into the fungal gardens. As the colony population enlarges, first large media workers and then major workers ("soldiers") are added. (Modified from Wilson, 1983b, in Wilson, 1985c; copyright 1985 American Association for the Advancement of Science.)*

Pheidole ants use fast feedback loops, the second category of superorganismic controls, in their division of labor. So long as the relative frequencies of adult minors and majors are near the species norm, the majors remain inactive most of the time. When they do work, they perform very specialized tasks in defense, cutting up large prey items, or milling seeds. The minors, in contrast, are in a virtually constant state of activity. They perform virtually all of the nonreproductive tasks of the colony, including food harvesting, nest construction, and care of the young. However, when the ratio of minors to majors is experimentally lowered to below 1:1 (from 3:1 to 20:1, the usual ratios according to species), so that minors are much scarcer than normal, the majors dramatically increase their repertory. Within an hour they start nursing larvae and transporting nest material, and in general behave like clumsy, overweight minor workers. They restore about 75% of the missing minors' activity, and they are able to rear larvae and pupae to maturity, feed the queen, and in general hold the colony together until the minor force is brought up to near normal levels (Wilson, 1984).

The feedback response that keeps the *Pheidole* division of labor in balance turns out to be a ritualized form of aversion, at least in the case of the South American species *P. pubiventris* (Wilson, 1985a; see Fig. 4). Major workers simply avoid the brood piles when substantial numbers of minor workers are on them; they turn away from the piles and walk off in another direction. When the number of minors is sufficiently reduced, the majors infiltrate the piles and commence brood care. This entrée is apparently the trigger for an overall increase in social behavior, measured in behavioral acts per individual major per hour, of approximately tenfold. Even the rate at which the majors groom themselves increases by this amount.

All of the *Pheidole* workers lack ovaries and hence are sterile. Their genetic fitness depends entirely on their ability to nurture the queen and rear her offspring, in other words their sisters and brothers. Natural selection occurs at the colony level, and a premium is placed on homeostatic controls that sustain optimal or near-optimal colony functioning through environmental perturbations. It is convenient to distinguish two kinds of flexibility on the part of ants and other kinds of social insects, according to level of organization (Wilson, 1985b). Responses in both categories contribute to social homeostasis. A colony as a whole shows a certain degree of resiliency in each task, as for example the ability of *Pheidole* to restore caste ratios and allocation of labor. Colony-level resiliency depends in turn on the degree

Figure 4. *Portion of a colony of the South American ant* Pheidole pubiventris. *Small-headed minor workers and large-headed major workers are deployed over a brood pile consisting of larvae and pupae. (From Wilson, 1985b.)*

of individual behavioral elasticity, which is defined as the amount by which the repertory of individual workers can be modified to restore the *status quo ante* when the caste ratios or other social processes are perturbed. The colonies of *Pheidole* show a remarkable resiliency due to the even more remarkable degree of elasticity in the behavior of the major workers.

Conclusion

I believe that Whitman, who stressed the stereotypy of animal behavior and the unity of biology, and Wheeler, who showed such foresight in formalizing the levels of organization, would have been quite comfortable with the current status of sociobiology. The cycle of organisms into superorganisms into organisms has proved conceptually sound and heuristic. Moreover, researchers have passed well beyond the mere listing of definitions and analogies when making comparisons of organisms and societies. In the case of social insects, with which I have dwelt primarily here, the aim of much of contemporary research is a detailed understanding of caste and division of labor, the modes of communication by which these processes are regulated, and the manner in which special patterns fit particular species to the idiosyncrasies of their respective environments. The long-range goal is the revelation of more general and heuristic principles by the meshing of comparable information from developmental biology and sociobiology. The central process at the level of the organism is morphogenesis, the programs by which individual cells or cell populations undergo changes in shape and position incident to organismic development. The central process at the level of the colony is sociogenesis, the programs by which individuals undergo changes in caste, behavior, and physical location incident to colonial development. Biology will continue to benefit by pressing a closer examination of the similarities between morphogenesis and sociogenesis.

LITERATURE CITED

Bayer, F. M. 1973. Colonial organization in octocorals. Pp. 69-93 in *Animal Colonies: Development and Function Through Time*, R. S. Boardman, A. H. Cheetham, and W. A. Oliver, Jr., eds. Dowden, Hutchinson, and Ross, Stroudsburg, PA.

Johnston, A., and E. O. Wilson. 1984. Correlates of variation in the major/minor ratio of the ant *Pheidole dentata* (Hymenoptera: Formicidae). *Ann. Entomol. Soc. Am.* **78:** 8-11.

Margulis, L. 1981. *Symbiosis in Cell Evolution.* W. H. Freeman, San Francisco, CA.

Oster, G. F., and E. O. Wilson. 1978. *Caste and Ecology in the Social Insects*. Princeton University Press, Princeton, NJ.

Wheeler, D. E., and H. F. Nijhout. 1983. Soldier determination in *Pheidole bicarinata*: effect of methoprene on caste and size within castes. *J. Ins. Physiol.* **29:** 847-854.

Wheeler, D. E., and H. F. Nijhout. 1984. Soldier determination in *Pheidole bicarinata*: inhibition by adult soldiers. *J. Ins. Physiol.* **30**: 127-135.

Wheeler, W. M. 1911. The ant colony as an organism. *J. Morphol.* **22(2):** 307-325.

Wheeler, W. M. 1927. *Emergent Evolution and the Social.* Kegan Paul, Trench, Trubner and Company, London.

Wheeler, W. M. 1928. *The Social Insects: Their Origin and Evolution.* Kegan Paul, Trench, Trubner and Company, London.

Wilson, E. O. 1968. The ergonomics of caste in the social insects. *Am. Nat.* **102:** 41-66.

Wilson, E. O. 1971. *The Insect Societies*. Belknap Press of Harvard University Press, Cambridge, MA.

Wilson, E. O. 1980. Caste and division of labor in leaf-cutter ants (Hymenoptera: Formicidae: *Atta*). II. The ergonomic optimization of leaf cutting. *Behav. Ecol. Sociobiol.* **7:** 157-165.

Wilson, E. O. 1983a. Caste and division of labor in leaf-cutter ants (Hymenoptera: Formicidae: *Atta*). III. Ergonomic resiliency in foraging by *A. cephalotes. Behav. Ecol. Sociobiol.* **14:** 47-54.

Wilson, E. O. 1983b. Caste and division of labor in leaf-cutter ants (Hymenoptera: Formicidae: *Atta*). IV. Colony ontogeny of *A. cephalotes*. *Behav. Ecol. Sociobiol.* **14:** 55-60.

Wilson, E. O. 1984. The relation between caste ratios and division of labor in the ant genus *Pheidole* (Hymenoptera: Formicidae). *Behav. Ecol. Sociobiol.* **16:** 89-98.

Wilson, E. O. 1985a. Between-caste aversion as a basis for division of labor in the ant *Pheidole pubiventris* (Hymenoptera: Formicidae). *Behav. Ecol. Sociobiol.* **17:** 35-37.

Wilson, E. O. 1985b. The principles of caste evolution. Pp. 307-324 in *Experimental Behavioral Ecology and Sociobiology,* B. Hölldobler and M. Lindauer, eds. G. Fischer Verlag, New York.

Wilson, E. O. 1985c. The sociogenesis of insect colonies. *Science* **228:** 1489-1495.

Wilson, E. O., and B. Hölldobler. 1988. Dense heterarchies and mass communication as the basis of organization in ant colonies. *TREE* **3:** 65-68.

Spencer F. Baird
(1823 - 1887)

Henry Bigelow
(1879 - 1967)

Alfred C. Redfield
(1890 - 1983)

Ecology in Woods Hole: Baird, Bigelow, and Redfield

Introduction by John H. Steele

In the last few years I have taken part in several celebrations of the centenaries of marine laboratories in Europe and the U.S. Many marine laboratories were founded throughout the world towards the end of the 19th century. There were, I believe, two spiritual parents for this series of births: the Naples lab in Italy and the Challenger expedition, both of which began in 1872. Together they provided a spectrum of interests from marine physiology to the study of the world's oceans. The study of fish, their evolution, their breeding cycles, and the causes of population fluctuations fitted naturally into this broad spectrum of interests in the ocean, from physiology to physical oceanography. This was certainly the range of interests of Spencer Baird and others who founded the fisheries lab here in Woods Hole. This interaction among the different components of marine science continued into the twentieth century and is represented here in Woods Hole by the cooperation between scientists who founded the Woods Hole Oceanographic Institution and those who worked at the National Marine Fisheries Service Laboratory and the Marine Biological Laboratory. In particular, the first director of the Woods Hole Oceanographic, Henry Bigelow, worked closely with other scientists, such as Bill Schroeder, in extending our knowledge of the fish populations in the northwest Atlantic.

In the late 1930s, however, and in the period immediately after World War II there was a separation of physiology, oceanography, and what we learned to call fisheries science. The reasons for this are of great interest, not only for the intellectual history of Woods Hole, but for an understanding of the way in which our science has developed. Part of the divergence may be linked to the social and economic climate, where the recession in the 1930s forced some of the components of marine science to concentrate in specific areas. But part of it was concerned with the separate intellectual development of each field.

It is a basic technique in science to attempt to consider the effects of variables one at a time. In the context of marine fisheries during the 1930s, 1940s, and 1950s, the greatest variable certainly was man's impact on fish populations. Thus the essential theoretical problem was to consider that variable and to assume that other factors were relatively constant. And so the mathematical models of fish populations took the ocean environment as a steady state. By focussing on the fish/man interaction they produced ideas that were of direct application to the problems of fisheries management and led the way in developing quantitative approaches in ecology. Of necessity they effectively eliminated the questions of the interaction between these fish stocks and their variable natural environment. At the same time there was a need in oceanography to develop theoretical concepts about the dynamics of the physical system and the chemical and biological processes. These would bring together the descriptions of the ocean structure, and of the populations of plankton in particular, living in the ocean. So, in turn, the oceanographers first looked for simplified models that could capture the equilibrium state of the ocean and ignore the variability. Once again, this was a very effective tool for intellectual development as well as for providing a basis for bringing together ideas about the physics, chemistry, and biology of the basic ocean processes. Here in Woods Hole this work is typified by the studies of Henry Stommel on ocean physics and by the work of Alfred Redfield on ocean chemistry and plankton dynamics.

In the last twenty years, however, and in the last decade in particular, these two separate systems of fisheries and ocean processes have been found to be insufficient to explain new observations of both the detail of distributions of organisms in the oceans and also of the longer-term trends in populations.

Changes in fish stocks and other communities may be not only an indication of man's direct involvement with the ocean, but also, less directly, an index of the response of the oceans to the changes in climate which are now such a critical element in the future development of industrial societies. We have come to realize that the oceans are the flywheel that control our climate, but it is an irregular flywheel that can change naturally as well as respond to man's intervention in the world's climate. But these interactions between ocean physics and marine communities require a much better and more general knowledge of the physiological and behavioral responses to environmental change. Thus the third strand from the beginnings of marine research a century ago—physiology—is now being woven back into the present fabric of our research. Here in Woods Hole this brings the MBL, the Oceanographic, and the fisheries lab back into a context reminiscent of the origins of this field and of **Spencer Baird, Henry Bigelow** and **Alfred Redfield**, three figures remembered in the upcoming chapter.

John Hobbie is a co-director of the Ecosystems Center at the Marine Biological Laboratory; he has also served as a member of the Marine Biological Laboratory's Board of Trustees. Hobbie, who specializes in the role of microbes in ecosystem processes, received his B.A. in zoology in 1957 from Dartmouth, his M.A. in zoology in 1959 from the University of California, Berkeley, and his Ph.D. from Indiana University in 1962. He worked as a post-doctoral fellow in the Antarctic and at Uppsala University in Sweden, where he became interested in microbial cycling of organic compounds in lakes. After serving on the faculty at North Carolina State University for ten years, Hobbie joined the staff of the Ecosystems Center as a senior scientist in 1976 and became the director in 1985. His research interests include arctic ecology and ecology of aquatic microbes.

Hobbie is the current president of the Association of Ecosystem Research Centers, a past president of the American Society of Limnology and Oceanography, a member of the Ecological Society of America and the 1988-1989 recipient of the Tage Erlander Visiting Professorship, awarded by the Swedish Natural Science Research Council.

John B. Pearce, Deputy Director of the Northeast Fisheries Science Center and Officer-in-Charge, NMFS Woods Hole Laboratory, conducts research concerned with the effects of contaminants on benthic community structure as well as studies on the interactions between parasitic crabs and their bivalve and tunicate hosts. He received his B.A. in biology in 1957 from Humboldt State University, and his M.A. in zoology (1960), and his Ph.D. (1962) from the University of Washington. He continued his research with benthic organisms as a post-doctoral fellow at the Scottish Marine Biological Association, Milport, Scotland, and the University of Copenhagen

Marine Lab, Helsingør, Denmark. He subsequently spent two years as a post-doctoral fellow at the Systematics Ecology Program, Marine Biological Laboratory, Woods Hole.

He later taught in the California university system and had adjunct appointments at the City University of New York and Lehigh and Rutgers Universities. He was a member of the United Nations Environmental Programme, Working Group 26, concerned with drafting the second Report on the "State of the World's Oceans" and also was Chairperson for the New Jersey Governor's Blue Ribbon Panel on Coastal Marine Pollution. He has served as a member of the Board of the Gulf of Maine Regional Marine Research Board. He has recently been nominated as a fellow of the American Association for the Advancement of Science and holds the Department of Commerce Gold Medal Award.

Ecology in Woods Hole: Baird, Bigelow, and Redfield

John E. Hobbie[1] and John B. Pearce[2]

[1]*Marine Biological Laboratory and*
[2]*National Marine Fisheries Science Center*

ECOLOGY, OR HOW ORGANISMS RELATE to their biological, physical, and chemical environment, is studied through investigations of individuals, populations, communities, and whole systems. Hundreds of ecologists working at all four of these levels of organization have worked at Woods Hole laboratories over the past century. In this chapter we are concerned with research at one of these levels and follow a single strand of the many possible connective threads. This trail leads from Spencer Baird's questions of the 1870s about the causes of the decline of fisheries to the large oceanographic and terrestrial projects of the 1980s and '90s at the Woods Hole Oceanographic Institution (WHOI), the Marine Biological Laboratory (MBL), and the National Marine Fisheries Research Center where questions about the controls of algal growth in the sea and global nitrogen cycling, for instance, have been investigated.

We describe here the questions, the research approaches, and the contributions of individual scientists rather than the current understanding of an ecological system or processes. We must do this because the literature on any given ecological topic is enormous and our theme is, in large part, historical. For example, the recent multi-authored description of Georges Bank, a nearby fishing ground, was a 5 kg book with 650 pages and 150 authors (Backus 1987).

Spencer Fullerton Baird—The Beginnings of Science in Woods Hole

One fundamental question, with great practical applications, is what controls the abundance of commercial fish, which are generally the top predators of the oceanic food chain. This question underlies a whole applied field of ecology—fisheries science. One pioneer in this field, Spencer F. Baird, visited Woods Hole in 1871 (Bourne, 1983) on an official mission to inquire into the decline of certain fisheries. Baird was not only a tireless promoter of science as Assistant Secretary of the Smithsonian Institution, but was also a prolific and talented researcher with early contacts with Audubon and numerous publications on birds, mammals, and reptiles. He had convinced Congress to create a U.S. Commission of Fish and Fisheries with an initial appropriation of $5000. As a result of his efforts, the Fisheries research station was created in 1882. Baird chose the Woods Hole site for his new station because of its closeness to both northern and southern marine faunas, its relatively pristine, unpolluted waters, and its access to good transportation with major urban centers.

Baird's early entrepreneurial and organizational efforts led over the decades to the founding of the MBL and WHOI in Woods Hole, but the scientific directions he established were, perhaps, more important. He insisted that the research of the Commission extend beyond determining the stocks of commercial fishes to investigate a number of possible causal factors for the decline of the catch of fishes; these factors included predation, environmental conditions, availability of prey and plant forage, migratory patterns, overfishing, and pollution. In other words, he and his co-workers took a holistic, ecological view of the problem. As a part of his goal he said that "...the relations existing between the fishes and the lower animals which serve as food for them were to be constantly kept in mind...." (Verrill and Smith, 1874; see also Galtsoff, 1962, and Dall, 1915). Given the state of measurement of marine productivity, the food chain, and ocean physics and chemistry in the 1880s and the early decades of the twentieth century, it is no wonder that it has taken a century for many of these factors to be adequately studied from the viewpoint of their effect on fish.

Another characteristic of Baird's approach was his persistence in involving academic scientists in early marine research. Many visiting scientists (often 15 or more) came to work with Baird during the summers. For example, Addison Verrill of Yale identified the organisms from 3000 dredge hauls to establish the species distribution of benthic animals in New England waters. He went beyond identification and stated "...we soon found that there are...three quite distinct assemblages of animal life, which are dependent upon and limited by definite physical conditions..." (Verrill and Smith, 1874). Thus, Verrill was one of the first to group benthic animals into assemblages and communities according to their habitat and to attempt to link biology and the physical sciences. One of his early assistants, E. B. Wilson, later was one of the founders of the MBL.

Henry Bigelow—Integrating Physics, Chemistry, and Biology

A principal beneficiary of the broad vision of Baird, a Harvard scientist Henry Bigelow, used the early Fisheries research ships for his pioneering oceanographic study of the Gulf of Maine. He was a protégé of Alexander Agassiz, trained as a systematic zoologist (coelenterates), accompanied Agassiz on expeditions to the east and tropical Pacific and to the Maldive Islands, and worked at the Harvard Museum of Comparative Zoology (MCZ). His achievement in the Gulf of Maine was not only to describe the floating (planktonic) life, but also to connect the seasonal and spatial distributions to the changes in chemistry and physics. He wrote (Bigelow, 1926):

> "Few living zoologists have been as fortunately placed as were we on setting sail on the Grampus from Gloucester on our first oceanographic cruise in the Gulf of Maine on July 9, 1912, for a veritable *mare incognitum* lay before us, so far as floating life was concerned...everything was yet to be learned as to what groups or species would prove predominant in the geographic and bathymetric variations; their seasonal successions, migrations and annual fluctuations; their temperature affinities, whether arctic boreal, or tropic; and whether they were oceanic or creatures of the coastal zone."

But how were these many factors to be studied in the ocean? Starting with the advice of Sir John Murray (of *Challenger* fame) and the Norwegian oceanographer Hjort, Bigelow developed the technique (described by Graham, 1968) of close observation and measurements at sea of physical, chemical, and biological factors. His Gulf of Maine study began in 1912 and was carried out over 12 years during a series of joint MCZ and U.S. Bureau of Fisheries cruises; the cruises covered all seasons of the year. Many of the observations were made from the schooner *Grampus* and other aging and "tipsy" ships (Bourne, 1983); Bigelow did most of the scientific work himself. His classic study, reported in book-sized monographs dealing with the

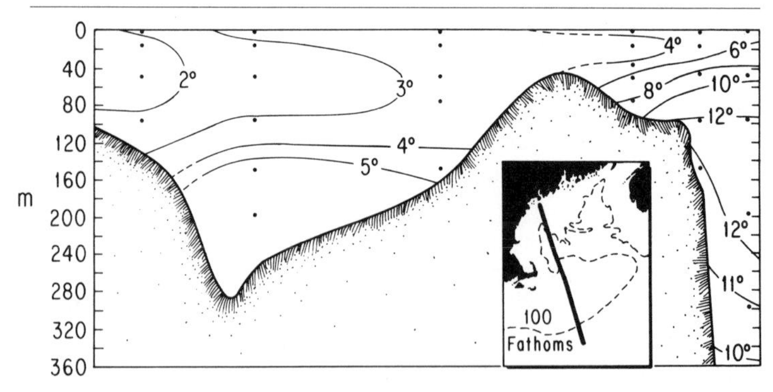

Figure 1. *Temperature profile of the Gulf of Maine across Georges Bank to the continental slope, 22 February to 4 March 1920 (from Bigelow, 1927).*

fishes, the plankton, and the circulation of the Gulf, has been called "...the first truly oceanographic study to be completed in North American waters" (Revelle, 1980).

The temperature data collected by Bigelow is typified by the late winter cruise of 1920 (Fig. 1). From data like these, and measurements of salinity plus drift bottles, he developed his ideas about the circulation in the Gulf (Fig. 2). This circulation scheme has helped scientists to explain the distribution patterns of organisms like larval fish and other meroplankton and the different stages of zooplankton (the haloplankton).

Despite his goal of going beyond the descriptive aspects of ecological science, Bigelow was unable to explain adequately much of the biological change that he documented so well. His explanations were qualitative, dictated by the techniques of his time. He did understand a lot about the causes of the spring diatom bloom in the open Gulf, but was puzzled by the events on Georges Bank (Bigelow, 1926; p. 484):

> "It is in just such areas as the open basin of the Gulf of Maine, where the transition from a state of free vertical circulation in early spring is sudden to one of very pronounced vertical stability in summer, when the supply of nitrogen and of phosphates from the deeps is thereby prevented, and where the silica content of the water is probably low...that the vernal flowerings of diatoms are briefest and vanish most completely after their culmination.
> The case is quite otherwise on Georges Bank, where one diatom

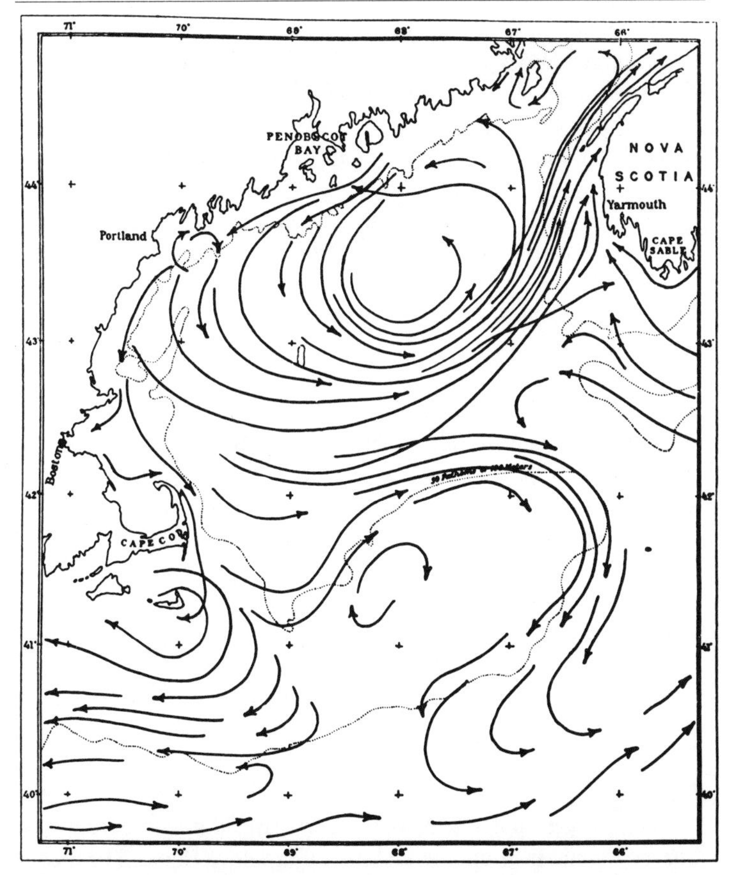

Figure 2. *The nontidal circulation of the Gulf of Maine, July to August (redrawn from Bigelow, 1927).*

community or another flourishes from late winter to midsummer, but where these flowerings are local by contrast to the extensive vernal flowerings in the inner part of the gulf.

The distance of the bank out from the land and the general distribution of salinity in the gulf forbid the possibility that the nutrients on which its diatoms depend are contributed directly by river water, while hydrography in general equally rules out any possible updraught of nutrients from the ocean deeps, this not being an area of upwelling."

We know now that the factors controlling any group of organisms—phytoplankton, zooplankton, or fish—interact in often complex ways that were not understood 60 years ago and are still not fully understood today. Also, the factors themselves change at time scales of days and weeks, not the months that were Bigelow's sampling period.

Bigelow's Gulf of Maine study ended in 1924, the same year that the aging fisheries vessel *Albatross* was sold and not replaced. "Bigelow, the leading American oceanographer on the east coast, was without the means to go to sea" (Burstyn, 1980). In 1927 a National Academy of Sciences committee—chaired by F. R. Lillie—was appointed to consider the role of the U.S. in oceanography. As secretary to the committee, Dr. Bigelow made the investigations and prepared the persuasive report that led to the founding of the Woods Hole Oceanographic Institution in 1930. As WHOI's first director, Bigelow set the pattern of bringing together the diverse skills of physical chemists, meteorologists, physiologists, bacteriologists, and algologists to study the sea. As noted by Graham (1968), Bigelow was "one of the founders of the new oceanography, that is oceanography with an ecological aim, so that instead of the mere description of what there was in the sea, there should be an explanation of the interconnections based on full knowledge and the applications of other branches of science."

Alfred Redfield—Finding Patterns

One of the first scientists hired by Bigelow to staff the new Oceanographic Institution was Alfred Redfield, a Harvard Professor of Physiology and long-term MBL scientist. From 1930 until 1942, he continued to devote his summers to research in Woods Hole and his winters to research and teaching at Harvard. In 1942 he was appointed Associate Director of WHOI and continued in this post until his retirement in 1957. Redfield combined singularly many of the "varied specialties which unite to increase our understanding of the aquatic environment" (Ketchum, 1965, p. R3). He described the relationship between circulation in the Gulf of Maine and the drift, distribution, and development of planktonic populations. He published on tidal effects in narrow embayments, harbors, and estuaries, and on deuterium as a tracer for fresh and salt water. In his later years he produced major works on the biology, geology, chemistry, and physics of salt marshes.

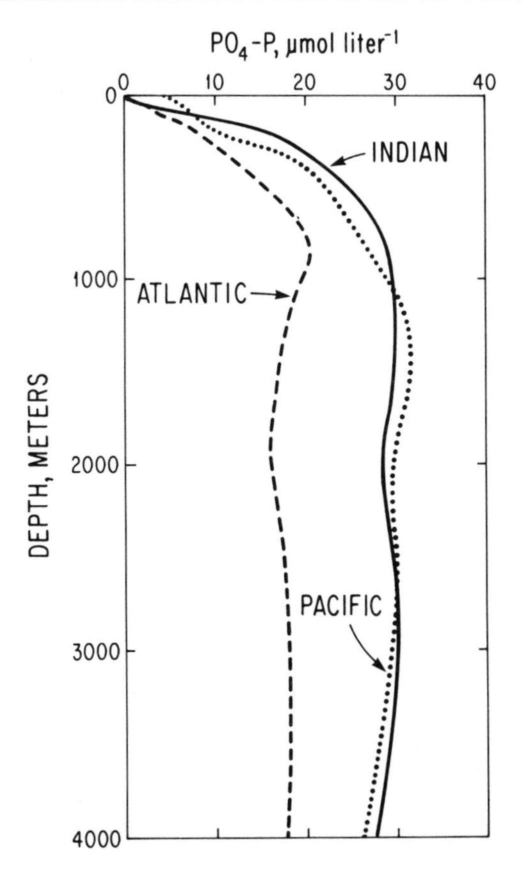

Figure 3. *Concentrations of phosphorus in various oceans of the world (redrawn from Sverdrup* et al., *1942).*

His unique contributions to ecology evolved from his early studies on the character of blood. One of his last papers on blood, in the *Quarterly Review of Biology*, dealt with the evolution of the respiratory function of blood (Redfield, 1933). While preparing this review, he was also writing a seminal paper on the chemical composition of organisms and seawater (Redfield, 1934). As Ketchum (1965) stated "...Dr. Redfield considered the ocean as a vast living organism in which the seawater serves as the 'blood stream' so that changes in its chemistry reflect the multitude of biological processes going on within it." Interestingly, in the late twentieth century many principal scientists are regarding large systems in the same light.

Redfield's contribution, however, went beyond a mere point of view; he was able to find a pattern in chemical data on nutrient concentrations in the sea. His method of finding the pattern was the same budget approach a physiologist would use to examine the general metabolism of an individual organism (Redfield, 1958). He first integrated observational data from many locations in the ocean and found that phosphate concentrations, for example, varied by more than six-fold world-wide (Fig. 3). Redfield observed that the variation was systematic and proceeded to look for the "world-wide mechanism at work" (Redfield, 1958). The mechanism turned out to be the cycling of carbon (C), nitrogen (N), and phosphorus (P) in planktonic organisms. When, however, he examined the nutrients available in deep water throughout the world, he found a constant ratio (Table I). Furthermore, the concentration differences between the surface waters and the deep waters had the same ratio as did the plankton itself (Table I). Redfield postulated that N, P, and C were fixed into phytoplankton during photosynthesis in the surface waters and later transported down from the surface waters by the settling of dead plankton. These elements were liberated from the plankton by decomposition in the deep waters, eventually to be returned to the surface by upwelling and turbulent mixing.

Table I

Atomic ratios of elements in the biochemical cycle (modified from Redfield, 1958)

	P	N	C
Analyses of plankton in surface waters	1	16	106
Changes in seawater, deep minus surface	1	15	105
Available in deep seawater	1	15	1000

Moreover, this pattern of the stoichiometry of nutrient concentrations and biological incorporation in the sea, now referred to as the Redfield Ratio, is a tool for predicting what should be happening in the sea as well as a meterstick for measuring the exceptions that must be explained and further investigated. From these relationships one can surmise that the carbonate and bicarbonate in the deep water (Table I) is about ten times more abundant than is needed for photosynthesis. Therefore, inorganic carbon is never limiting for primary production in the sea. Redfield (1958) explained some of the consequences of this ratio for our understanding of other possible limiting elements: "In a similar way sulfur, an important plant nutrient, is available in great quantities relative to phosphorus and nitrogen and the same is true for calcium, magnesium, and potassium."

Phosphorus and nitrogen thus appear to be the constituents of seawater present in limiting quantities. Although this was alluded to when Harvey (1926) pointed out "the remarkable fact that in the English Channel plant growth should strip sea water of both nitrates and phosphates at about the same time," it was Redfield who recognized the pattern and linked the disappearance to biological mechanisms and to the world ocean.

Redfield's Legacy in Woods Hole—A Global View

The idea that biological processes reciprocally affect the chemistry of the water and air is one of the cornerstones of the modern field of biogeochemistry. One way to investigate and verify further these relationships is based upon Redfield's ideas of element matching in biological systems. That is, the important elements controlling many aspects of carbon cycling are also among the major elements making up the biota of the earth; the concentrations of these important nutrient elements are linked according to the ratios laid out in the Redfield Ratio. A concrete example of the value of this approach comes from work on the global cycling of carbon carried out at the MBL's Ecosystems Center.

The example, published by Bruce Peterson and Jerry Melillo in 1985, is summarized as follows. Carbon dioxide is increasing in the atmosphere and will likely increase the temperature of the earth through the greenhouse effect. It has been hypothesized that the rate of carbon dioxide buildup is reduced significantly by increases in the long-term storage of carbon in the open ocean, coastal ocean, and forests caused by the nitrogen and phosphorus that is added by human activity (called eutrophication). Because carbon is stored in the forests and in the ocean sediments as organic matter, it must always be accompanied by N and P in quantities dictated by the various element ratios of the different types of organic matter. Therefore, the N and P supply limits the amount of "storage" possible. It is calculated that the buildup of carbon dioxide in the atmosphere is reduced by only 3% or less each year as a result of N and P eutrophication, and the authors conclude that the increases in the storage of carbon are trivial relative to the increases in the release of carbon from the combustion of fossil fuel.

The 1985 paper was concerned with the global cycle and how to explain certain apparent discrepancies. This cycle was explained from the biologist's viewpoint by Hobbie *et al.* (1984):

SOURCES [fossil fuel + harvest and clearing] EQUAL
SINKS [atmospheric increase + ocean uptake + others by difference]

"SOURCE" refers to carbon dioxide that enters the atmosphere in excess of the natural (*i.e.*, pre-industrial revolution) sources. "SINK" refers to the fate of that excess carbon dioxide. Sources must equal sinks, but the equation cannot be balanced at the present time because only two of the terms have been measured: the fossil fuel release rates and the rate of increase of carbon dioxide in the atmosphere. Chemical oceanographers believe that they can estimate the ocean uptake of carbon, and ecologists believe that they can estimate the release due to harvest and clearing. However, when either of these rates are put into the equation it does not balance. As a result, the hypothesis has been put forward, mainly by chemical oceanographers (Hobbie *et al.*, 1984), that there must be an important biotic sink in the category of "others" and that it is to be found in the forests and in the sediments and particulate matter of oceans and coastal oceans. Peterson and Melillo (1985) attempted to place some logical limits on the amount of storage that can take place to find out just how important this "other" category may be.

To understand the ideas described here, it must be clear that we are talking about the sources and sinks of the excess CO_2; that is, the carbon that is added to the atmosphere by human activities. This is a small amount relative to the total amount present in the atmosphere and cycling each year, but it is the amount that results in the gradual year-by-year buildup of CO_2 in the atmosphere.

The annual fluxes and Redfield Ratios in the global circulation of carbon for 1980 are given in Table II. Note that Peterson and Melillo (1985) used the term "elemental ratios" instead of "Redfield Ratios" because Redfield wrote only about plankton. One conclusion from this table is that the total net photosynthesis of the terrestrial and aquatic systems is large relative to the rate of carbon release from fossil fuel; therefore, there is the capacity for biotic storage to be an appreciable sink for carbon. A second conclusion is that the combustion of fossil fuels releases about half the amount of nitrogen to the atmosphere as is produced by the fertilizer industry. A third conclusion is that the Redfield Ratio in fossil fuel and in terrestrial vegetation is high relative to the ratios in soils, plankton, and sediment.

Table II

Annual fluxes and Redfield Ratios for 1980 (summarized from Peterson and Melillo, 1985). Fluxes are in teragrams (10^{12} g) of carbon and ratios are by weight

Process	Flux ($Tg\ yr^{-1}$)	Redfield Ratio C:N:P
Terrestrial net photosynthesis	55,000 carbon	
Vegetation		800:9:1
Soils		120:9:1
Aquatic net photosynthesis	30,000 carbon	
Plankton		40:7:1
Sediments		80:7:1
Fossil fuel combustion	5,200 carbon 20 nitrogen	9,000:36:1
Fertilizer production	59 nitrogen 12 phosphorus	

Other evidence summarized in this 1985 paper allows the conclusion that photosynthesis and storage of terrestrial and aquatic ecosystems can be stimulated by additional nitrogen and phosphorus but not by additional carbon dioxide. The storage is not of carbon alone but of organic matter that is made up of C, N, and P in well-known ratios: there must be element matching. This is a crucial point for the entire argument because it means that a potential increase in the amount of carbon stored in the biosphere can be calculated from our knowledge of the increase in the amount of N and P available.

From the data in Table II, a series of calculations may be made of the change to be expected in the various biotic reservoirs (forests, coastal zone, ocean) in response to N and P added by human activities. The question asked is how much will these human-induced nutrient increases alter the distribution of carbon between the atmosphere and biosphere? In other words, how much will the carbon sinks of forest, coastal zone, and ocean be increased?

Forests and forest soils, which contain more than 90% of the terrestrial biosphere's carbon, were divided into regions (Table III) and into compartments of vegetation, litter, and soil. Each region was then assigned a reasonable amount of the total N released to the atmosphere from fossil fuels (it was assumed that 30% or 6 Tg reached the forests) to correct for the non-uniform distribution of fossil fuel use (temperate and tropical forests received 4 and 0.5 Tg, respectively). Finally, it was assumed that only 60% of the added nitrogen was retained and resulted in growth and storage. The potential total storage (Table III) is around 100 Tg C yr^{-1}, a large amount but still a change of only 1 part per 5000 of the total forest carbon. This small relative change is impossible to measure in a forest but may be estimated as above.

In the coastal zones of the oceans, the sediments bury some of the organic carbon carried by rivers to the ocean as well as some of the algal primary production. Both the riverine transport of organic carbon into the coastal zone and the rates of algal photosynthesis are now higher than in pre-industrial revolution times because of human activities including eutrophication. The amount of the sink (the excess over "natural" processes) may be calculated from the fluxes of C, N, and P to the coastal zone (Table IV, as documented by Meybeck, 1982). This increased burial will result from the human derived fluxes (Table IV) that increase algal photosynthesis. The flux of 7.0 Tg N and 1.0 Tg P is about at the Redfield Ratio. Therefore, as described in Table II, the increase in algal photosynthesis will be 40 to 80 Tg C yr^{-1}. This photosynthesis increase is added to the human derived inflow of carbon (100 Tg from Table IV) to give 140-180 Tg C yr^{-1} as the maximum potential burial. Peterson and Melillo (1985) discuss all the processes affecting this carbon and conclude that a reasonable upper limit for burial of the excess carbon is 50 Tg yr^{-1}.

Table III

Potential carbon storage in forests due to nitrogen loading from fossil fuel (modified from Peterson and Melillo, 1985)

Region and Area	Carbon (kg C m^{-2})	Compartments (% of N)	N load (Tg yr^{-1})	C storage (Tg yr^{-1})
Tropics	31.5		0.5	
2450 million ha		vegetation (14.8)		6.7
		litter (0.3)		0.1
		soil (84.8)		3.3
Temperate	29.1		4.0	
12 million ha		vegetation (10.3)		37.1
		litter (2.3)		2.8
		soil (87.4)		27.3
Boreal	26.5		1.0	
12 million ha		vegetation (5.4)		4.9
		litter (3.2)		1.0
		soil (91.5)		7.1
Woodland and				
shrubland	13.2		0.5	
8.5 million ha		vegetation (6.9)		3.2
		litter (1.0)		0.2
		soil (91.5)		3.6
Totals			6.0	97.0

Table IV

Fluxes of carbon, nitrogen, and phosphorus to the ocean in rivers (adapted from Meybeck, 1982)

Category	Natural flux ($Tg\ yr^{-1}$)	Human derived ($Tg\ yr^{-1}$)
Carbon		
dissolved organic carbon	215	
particulate organic carbon	180	
	—	
total	395	100
Nitrogen		
dissolved		
inorganic 4.25, organic 11.00	15.25	7.0
particulate	21.00	
	——	
total	36.25	
Phosphorus		
dissolved	1.00	1.0
particulate	20.00	
	——	
total	21.00	

In the open ocean, airborne nutrients from fossil fuel combustion are the only nutrients causing an increase in photosynthesis over the "natural" rate. This nutrient increase is about 30% of the 20 Tg N released from fossil fuel combustion (Table II) or 6 Tg N yr^{-1}. The resultant increase in photosynthesis of 40 Tg C (using the Redfield Ratio for plankton in Table II) may sink into and through the permanent thermocline and provide a temporary sink for carbon. This estimate may be too high, however, because of a lack of phosphorus in certain open ocean areas.

The sum of all the possible biotic sinks of atmospheric CO_2 reported here totals less than 200 Tg yr^{-1} (Table V). This is a small amount relative to the 5200 Tg released by fossil fuel combustion and leads to the conclusion that the rate of atmospheric CO_2 increase has not been significantly reduced by increased storage of carbon in the biotic sinks of forests, and coastal and open oceans. The reason for this is apparent after comparing the Redfield Ratios of the sources and sinks (Table II). As Peterson and Melillo state (1985, p. 125), "The burning of fossil fuels and forest harvest and clearing release much more carbon as CO_2 relative to N and P than can be sequestered as organic carbon with this same amount of nutrient in the biotic sinks. Even when other sources of N and P are considered, there is still a large surplus of CO_2." The insights of Alfred Redfield into the nutrient matching requirement of biological systems have provided us with an important legacy, indeed have given us a powerful tool for future analyses.

Table V

Potential sinks for the excess carbon in forests, sediments, and deep ocean water as estimated from nutrient matching (modified from Peterson and Melillo, 1985)

Sink for carbon	Amount (Tg yr^{-1})
Forest vegetation and soils due to N released from fossil fuels	100
Coastal sediments due to land runoff and eutrophication	50
Open ocean deep waters due to fossil fuel release	40
total	190

Bigelow's Methods Today—Georges Bank Studies

The ideas pioneered by Henry Bigelow about how to conduct oceanographic science and the need for multidisciplinary studies have evolved to a highly complex state. One result in the 1990s is the multinational interdisciplinary Joint Global Ocean Flux Study (JGOFS) the goals of which are:

1) To identify and quantify the physical, chemical, and biological processes controlling biogeochemical cycling in the ocean, and the interaction of these processes with the global atmosphere, and 2) to understand the processes governing the production and fate of biogenic materials in the sea well enough to predict their influences on, and responses to, global scale perturbations (Global Ocean Flux Study Committee, 1984).

An example of how physics, chemistry, and biology can combine to explain biological phenomena in the sea comes from a study of Georges Bank (Horne *et al.*, 1989), summarized below. The overall question addressed in this study is what controls and causes the extraordinary algal and fishery productivity of this fertile fishing area?

The authors state that the conceptual basis of the study is that algal production (photosynthesis) is controlled by N input from two sources (Dugdale and Goering, 1967). One input is "regenerated" N formed on Georges Bank (the study site) from the metabolism of plants, animals, and microbes. If this were the only input of N, an ecosystem would soon run down because there are constant losses through such processes as the harvesting of fish, settling of algae, and denitrification by microbes. The second input is "new" N from outside the system, such as from rain water (not important on Georges Bank) or from the deep oceanic waters rich in nitrate. The "regenerated" production is that required to support the metabolic needs of the system; the "new" production is the basis for the potential reproduction and growth of the community. From the viewpoint of any fishery or food chain, harvest cannot exceed new production without destroying the system. In fact, from this viewpoint the most important single thing to investigate is the rate of supply of the "new" nitrogen because it is this that ultimately provides the nutrients for the fishery harvest. For Georges Bank, a convenient fishery animal to discuss is the scallop, a filter feeder that depends upon algae and particulate matter that sinks from the surface layers. But first, let's consider the nitrogen requirements on Georges Bank. The source of any new nitrogen is nitrate from the deeper waters surrounding the Bank.

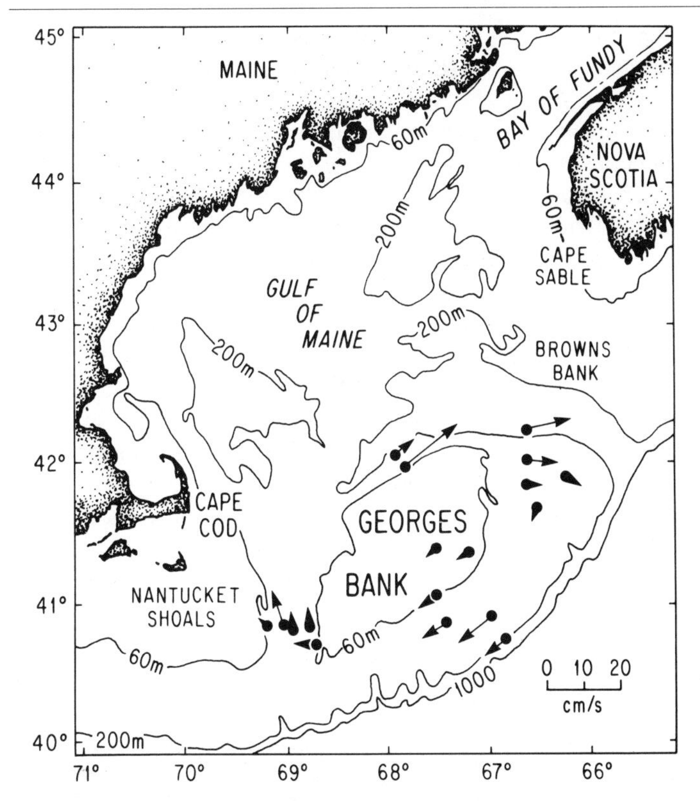

Figure 4. *Georges Bank location, water depths (meters), and mean currents (redrawn from Horne* et al., *1989).*

Georges Bank is a shallow submarine bank lying off Cape Cod between the shelf slope and Nova Scotia (Fig. 4). Because of its proximity to the Gulf of Maine-Bay of Fundy tidal system, the Bank has strong tidal currents of up to 1 m s^{-1}. These tides, plus seasonal density gradients, cause the persistent clockwise circulation (Fig. 2) deduced by Bigelow (1927). These strong tides also keep the water over the central Bank well-mixed vertically throughout the year. The surrounding waters, in contrast, are well-mixed during the winter but stratified during the spring and summer. For this reason, during the spring-fall period, a tidal front develops around the central Bank between the mixed and stratified waters. The front is often seen on summer satellite views of water temperature as a 10-40 km annulus of rapid temperature change between the warm waters of the Gulf and open ocean and the cooler surface water of the mixed central Bank areas. In Figure 5a, the well-mixed water on the central Bank is 14°C while the surrounding surface open ocean water is above 17°C and the underlying deep open ocean water is 6-9°C.

Horne *et al.* found that the front is also evident in the distribution of nitrate (Fig. 5b). Nitrate concentrations are highest in deep waters of the Gulf of Maine (up to 17 mmol liter^{-1}) and lowest (below 1 mmol liter^{-1}) at the surface of the Gulf of Maine and over the central Bank. In this figure the front lies between stations 306 and 308. An abrupt increase in the chlorophyll concentration, a measurement of the quantity of phytoplankton present, is also seen along the front in Figure 5d. Ammonium (Fig. 5c) is not a good measure of the location of the front because it is present everywhere at low concentrations and is rapidly biologically degraded, regenerated, and taken up by the phytoplankton.

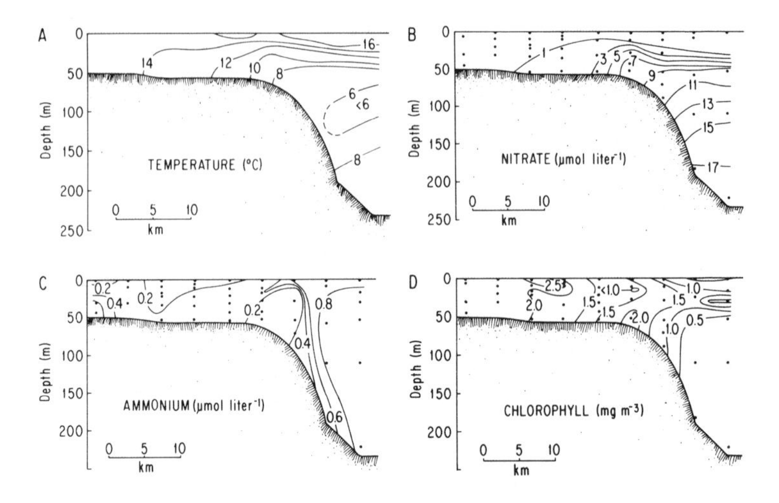

Figure 5. *Temperature (a, °C), nitrate concentrations (b, µmol liter^{-1}), ammonium concentrations (c, µmol liter^{-1}), and chlorophyll concentrations (d, mg m^{-3}) along a section of northwestern Georges Bank on 9-10 August 1985 (modified from Horne* et al., *1989).*

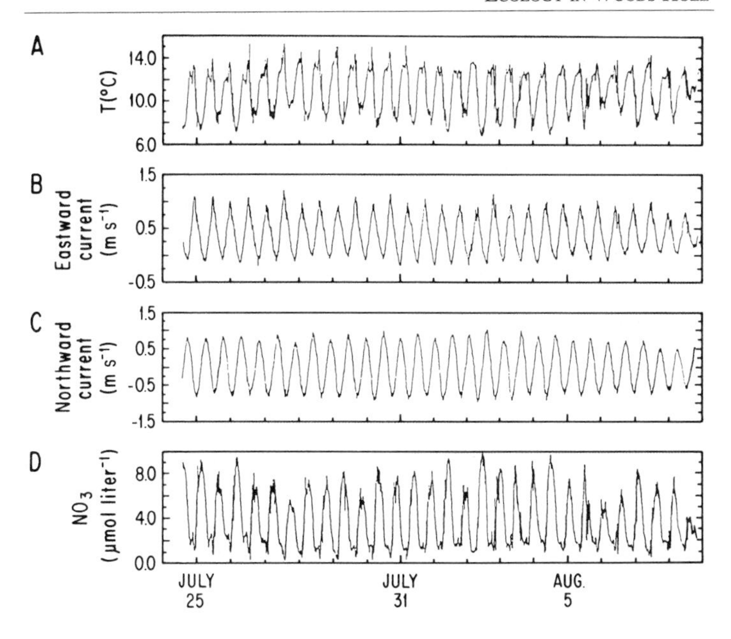

Figure 6. *Time series of moored measurements of (a) temperature (°C), (b) eastward current (along the front), (c) northward current (normal to the front), taken at 33 m in a water depth of 63 m on the northern side of Georges Bank. The nitrate time series (d) is calculated from the temperature data by using a nitrate-temperature relation as discussed in the text (redrawn from Horne* et al., *1989).*

It is important to recognize that the front is not stationary but advances and retreats with the tide. The picture of temperature and nutrients given in Figure 5 is, therefore, a snapshot at one stage of these tidally generated movements. The motion of the front is evident in a time series of temperature and current measurements taken from instruments moored at mid-depth (33 m below the surface) at a station along the northwestern edge of the Bank (Fig. 6). The oscillations of temperature are consistent with water parcel movements of about 10 km on each tide. These data from 33 m, and a similar set from 13 m, were used by Horne *et al.* to calculate that the along-front (eastward) component of mean current is 0.39 m s^{-1} and 0.30 m s^{-1} at 13 and 33 m levels, respectively, and the cross-frontal component is -0.05 m s^{-1} and 0.03 m s^{-1}, at the same depths, directed southward towards the center of the Bank. Thus, the clockwise rotary semidiurnal tidal current, plus an eastward current, dominates the current field at the northern side of the Bank.

To establish the nitrogen requirements of the phytoplankton on Georges Bank, Horne *et al.* (1989) measured rates of carbon and nitrogen incorporation in short-term bottle experiments in which $H^{14}CO_3^-$, $^{15}NO_3^-$, and $^{15}NH_4^+$ were added to plankton samples (Table VI). The central zone of Georges Bank has negligible nitrate, some ammonium, high phytoplankton biomass (chlorophyll), and low nitrate uptake rates while the zone at the edge of the Bank has much higher nitrate uptake. Ammonium uptake is more important than nitrate in the off-Bank and oceanic samples. Horne *et al.* reported that the nitrate demand of the phytoplankton at the time of sampling was 0.09 mmol m^{-2} s^{-1} in the central mixed zone, 0.36 mmol m^{-2} s^{-1} at the front, 0.18 mmol m^{-2} s^{-1} at the off-bank station, and 0.02 mmol m^{-2} s^{-1} at the oceanic stations (Table VII).

Table VI

Euphotic zone integrals of chlorophyll, dissolved nitrate, and ammonium, and uptake rates of carbon (photosynthesis), nitrate, and ammonium on Georges Bank and in the NW Atlantic (summarized from Horne et al., *1989). The depth of the euphotic zone is the depth of the 1% level of the photosynthetic active radiation*

Location	Depth (m)	Chl mg m^{-2}	Nutrients mmol m^{-2}		Uptake Rates mmol m^{-2} h^{-1}		
			NO_3	NH_4	C	NO_3	NH_4
Central Bank	35	135	0	4	230	0.33	0.94
Edge of Bank	42	97	95	8.5	223	1.0	0.91
Off-Bank	80	75	350	8	210	1.1	2.0
Oceanic	100	29	20	16	79	0.8	0.97

Table VII

Estimated nitrate demand of phytoplankton in the mixed and frontal regions of Georges Bank. Here "front" is the part (about half) of the frontal zone inside the line across which the supply rate has been estimated (extracted from Table 3 of Horne et al., *1989)*

	Front	Mixed	Total
Per unit area (mmol m^{-2} s^{-1})	0.36	0.09	
Per unit length of front (mmol m^{-2} s^{-1})	2.9	2.5	5.4

Now that the amount of nitrate needed in the systems is known, the question necessary to complete the understanding becomes the mechanism of the supply of nitrate and its control. Horne *et al.* (1989) point to the horizontal exchange across the tidal front as the most likely mechanism but state that how this particular mechanism actually works is an unresolved question in oceanography. They believe that the fluxes of heat and nitrate that they observed on Georges Bank are due mostly to "skew diffusion," which they explain by stating (Horne *et al.*, 1989), "In a wave or eddy field with a preferred sense of rotation, there can exist a component of the eddy scalar flux (called the skew flux) that is perpendicular to the mean scalar gradient (that is along the isolines). This phenomenon...contrasts with the better-known Fickian diffusion where the flux is downgradient..."

Luckily, in the case of the Georges Bank front, the actual heat transfer across the front could be measured by the techniques of moored instruments that produced the data shown in Figure 6. This works here because of the contrast between the cold water (7-9°C) off the Bank and the warm water (13-17°C) on the Bank (Fig. 5) and the resulting signal (Fig. 6) as these different waters move back and forth past the moored instrument. Horne *et al.* (1989) also measured the nitrate concentration in the water at the moored station and found that there was an inverse correlation between the nitrate concentration and the water temperature. This is also seen in Figure 5a and b where the warm Bank water contained less nitrate than the cold, deep-lying off-Bank water. The correlation was so good that they could calculate the nitrate in the tidal water moving back and forth across the Bank from the temperature alone, with an uncertainty of about ±0.5 μmol $liter^{-1}$ (Fig. 6d). The resulting cross-front nitrate transport is given in Table VIII. The units for the transport seem strange to a biologist but it is easiest to imagine that the front is a vertical surface, like a curtain, in the water. The cross-front transport rates are reported as amounts passing through a square meter of this "curtain." The total on-Bank nitrate transport was 0.01 mmol m^{-2} s^{-1} at 13 m and 0.28 mmol m^{-2} s^{-1} at 33 m. The authors then depth integrated these values and assumed that the 13 m value was typical of the top 22 m of the "curtain" and the 33 m value was typical of the bottom 44 m. The total nitrate transport can then be expressed with units of per meter of the front (= the "curtain"). Here it is around 12 mmol m^{-1} s^{-1}.

Table VIII

Cross-frontal nitrate fluxes (mmol m^{-2} s^{-1}) at two depths at the mooring site on the northern Georges Bank, 24 July-10 August 1985. Negative values indicate on-bank fluxes (modified from Horne et al., *1989)*

Eddy Flux	13 m	33 m
Covariance estimate	0.03	-0.16
Co-spectral estimate	0.03	-0.17
M_2-band estimate (harmonic analysis)	0.02	-0.13
Mean-flow flux	-0.04	-0.12
Total flux (eddy + mean)	-0.01	-0.28

This nitrate transport or supply has to support the new production in the frontal zone as well as in the mixed area. This new production consumes 5.4 mmol nitrate m^{-1} s^{-1} as described earlier (Table VII). Thus, Horne *et al.* (1989) conclude that the estimated nitrate supply is about twice the demand, an excellent level of agreement given the uncertainties in the calculations. In other words, they have suggested a mechanism that will supply the nitrate and have shown that it will supply as much or even more than the nitrate required. The authors suggest that it is this constant, year-round supply of nitrate from the deep water that accounts for the high productivity of Georges Bank.

The scallop (*Placopecten magellanicus*) is a filter-feeding benthic bivalve mollusk that depends on materials—mainly phytoplankton—sinking from the surface layers. They are, therefore, directly dependent upon new production and are an important fishery. Horne *et al.* (1989) point out that in the Canadian sector, the densest scallop concentrations form an annular pattern around the northern edge of the Bank and that this dense concentration corresponds with the general location of the frontal zone. In other words, scallop distribution is determined by the abundance of their planktonic food supply. Furthermore, the amount of new nitrogen that this population consumes is about 15 mol s^{-1}, or less than 1% of the 2000 mol N s^{-1} produced in photosynthesis during the summer and much less than the 5000 mol N s^{-1} supplied to this sector of the Bank by cross-frontal transfer.

Thus, Horne *et al.* (1989) believe that they have confirmed the proposed system to account for Georges Bank's high productivity in the cross-frontal nutrient flux driven by a "skew flux" associated with the tidal currents. Their approach was multidisciplinary, involving the close cooperation of physical, chemical, and biological oceanographers working at the same stations and at the same time. Their work is an excellent example of how Bigelow's ideas about how oceanographic problems should be approached have found their modern expression.

Baird's questions about the possible causes of the decline of certain fisheries required much more complex answers than he could have imagined a century ago. The road to the answers was charted by Bigelow, a pioneer in the multidisciplinary approach, and by Redfield, who discovered the quantitative link between algal growth and the nutrients. Modern studies, using the approaches of Bigelow and Redfield, are explaining how the nutrient supply rate drives the high algal productivity of systems such as Georges Bank and how nutrient supply rates may limit the amount of atmospheric carbon dioxide that can be sequestered in the biosphere.

Literature Cited

Backus, R. H. 1987. *Georges Bank.* Massachusetts Institute of Technology Press, Cambridge, MA.

Bigelow, H. B. 1926. Plankton of the offshore waters of the Gulf of Maine. *Bull. Bur. Fish.* **40**, 1924, part 2. 509 pp. U. S. Govt. Printing Office Doc. 968.

Bigelow, H. B. 1927. Physical oceanography of the Gulf of Maine. *Bull. U.S. Bur. Fish.* **40:** 511-1027.

Bourne, D. W. 1983. The Fisheries. Pp. 160-172 in *Woods Hole Reflections*, M. L. Smith, ed. Woods Hole Historical Collection, Woods Hole, MA.

Burstyn, H. L. 1980. Reviving American oceanography: Frank Lillie, Wickliffe Rose, and the founding of the Woods Hole Oceanographic Institution. Pp. 57-66 in *Oceanography: The Past*, M. Sears and D. Merriman, eds. Springer-Verlag, New York, Heidelberg, Berlin.

Dall, W. H. 1915. *Spencer Fullerton Baird, A Biography*. J. B. Lippincott Company, Philadelphia, PA.

Dugdale, R. C., and J. J. Goering. 1967. Uptake of new and regenerated forms of nitrogen in primary productivity. *Limnol. Oceanogr*. **12**: 196-206.

Galtsoff, P. S. 1962. The Story of the Bureau of Commercial Fisheries Biological Laboratory, Woods Hole, Massachusetts. United States Department of the Interior, Circular 145.

Global Ocean Flux Study Committee. 1984. Global Ocean Flux Study, Proceedings of a Workshop. National Academy Press, Washington, DC.

Graham, M. 1968. An appreciation of Henry Bryant Bigelow. *Deep Sea Res.* **15(2):** 125-132.

Harvey, H. W. 1926. Nitrate in the sea. *J. Mar. Biol. Assoc. U.K.* **14**: 71-88.

Hobbie, J. E., J. J. Cole, J. L. Dungan, R. A. Houghton, and B. J. Peterson. 1984. Role of biota in the CO_2 balance: the controversy. *BioScience* **34**: 492-498.

Horne, E. P. W., J. W. Loder, W. G. Harrison, R. Mohn, M. R. Lewis, B. Irwin, and T. Platt. 1989. Nitrate supply and demand at the Georges Bank tidal front. Topics in Marine Biology, J. D. Ros, ed. *Scient. Mar.* **53:** 145-158.

Ketchum, B. H. 1965. Forward to Alfred C. Redfield 75th Anniversary Volume. *Limnol. Oceanogr.* **10** (suppl.): R1-R8.

Meybeck, M. 1982. Carbon, nitrogen, and phosphorus transport by world rivers. *Am. J. Sci.* **202**: 401-450.

Peterson, B. J. and J. M. Melillo. 1985. The potential storage of carbon caused by eutrophication of the biosphere. *Tellus* **37B:** 117-127.

Redfield, A. C. 1933. The evolution of the respiratory function of the blood. *Q. Rev. Biol.* **8**: 31-57.

Redfield, A. C. 1934. On the proportions of organic derivatives in sea water and their relation to the composition of the plankton. Pp. 176-192 in *James Johnstone Memorial Volume.* University of Liverpool Press.

Redfield, A. C. 1958. The inadequacy of experiment in marine biology. Pp. 17-26 in *Perspectives in Marine Biology*, A. A. Buzzati-Traverso, ed. University of California Press, Berkeley and Los Angeles.

Revelle, R. 1980. The Oceanographic and how it grew. Pp. 10-24 in *Oceanography: The Past*, M. Sears and D. Merriman, eds. Springer-Verlag, New York, Heidleberg, and Berlin.

Sverdrup, H. U., M. W. Johnson, and R. H. Fleming. 1942. *The Oceans*. Prentice-Hall, Englewood Cliffs, NJ.

Verrill, A. E., and S. I. Smith. 1874. *The Invertebrate Animals of Vineyard Sound and the Adjacent Waters.* U. S. Government Printing Office, Washington, DC. 478 pp.

INDEX